MUSIC PRODUCTION:

RECORDING

MUSIC PRODUCTION: RECORDING

A Guide for Producers, Engineers and Musicians

Carlos Lellis Ferreira

Focal Press
Taylor & Francis Group

NEW YORK AND LONDON

First published 2013
by Focal Press
70 Blanchard Rd Suite 402
Burlington, MA 01803

Simultaneously published in the UK
by Focal Press
2 Park Square, Milton Park, Abingdon, Oxon OX14 4RN

Focal Press is an imprint of the Taylor & Francis Group, an informa business.

Library of Congress Cataloging in Publication Data

Ferreira, Carlos Lellis.
Music production : recording : a guide for producers, engineers and musicians / Carlos Lellis Ferreira.
pages cm
Includes bibliographical references.
1. Popular music--Production and direction. 2. Sound recordings--Production and direction. 3. Sound--Recording and reproducing. I. Title.
ML3790.F49 2013
781.49--dc23
2012046713

ISBN: 978-0-240-52273-9 (pbk)
ISBN: 978-0-240-52274-6 (ebk)

Typeset in Century Gothic and Helvetica Neue.

Cover courtesy of Stefano Battarola, Jessie Zarazaga and SAE London.

Printed and bound in India by Replika Press Pvt. Ltd.

CONTENTS

ACKNOWLEDGEMENTS

Thank you:

to Rebecca, Harley and Ella,

to the Ronchinis and the Lellis Ferreiras,

to Stefano, Jules and Jessie,

to all the musicians, engineers, producers and music industry professionals I have worked with,

to all my students past and present, for allowing me to learn with them.

NOTE

This book is comprised of different kinds of information. Some of it is what I believe recordists and producers need to know, while the rest is what I think they may choose to know. From my experience, knowledge commonly equates to confidence, a very valuable asset in music production.

The observations and suggestions presented here come from a lifetime in music. Whether performing, recording, mixing, mastering, producing or teaching I have been immersed in a musical universe from as early as I can remember.

Music has given me something to believe in and it has brought love into my life. If you feel as I do about it, then this book is dedicated to you.

MUSIC PRODUCTION

RECORDING

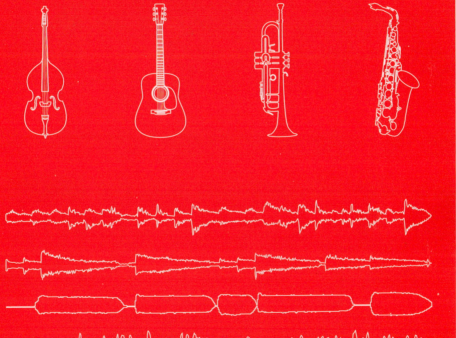

FOREWORD

FOREWORD

Soon after the author of this book and I met we both quickly realised that we shared the same passion and excitement for music and audio. Not the mad professor audio laboratory that is the mixing process, but the actual moment of audio creation, the recording process.

So how do we start in music recording? We have a drummer, sticks at the ready (silent!), guitarists have their first chord or note fretted, hand ready to strike strings, keyboard player fingers in chord shapes, ready to land on keys, vocalist, throat cleared thinking about the first note or perhaps a whole orchestra, silent, ready, watching for the conductor's signal to start. In the control room, the engineer's mind races through a list of all the potential problems encountered during set-up, have these all been addressed? Can the 'go ahead' be given to start? A mental note of LEDs and meter dials that need to be watched for overload or lack of signal is made. The technical set-up time is over. A new phase begins, one of hope for the performer's music to flow perfectly. Fingers are crossed for them and the technology. This is going to be it! This is going to be the one!

Recording!

Julian 'Jules' Standen
Gearslutz.com Audio Forum
London, October 2012

MUSIC PRODUCTION

RECORDING

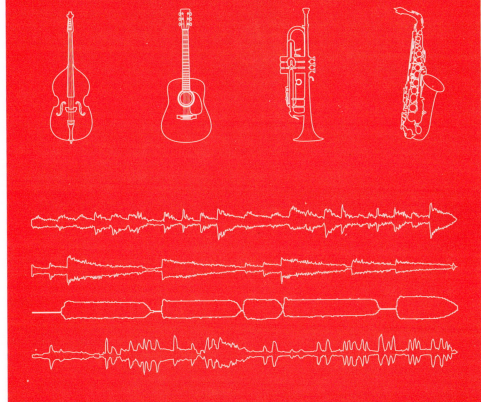

INTRODUCTION

THE ACT OF RECORDING

The year of 2012 marked the 135th anniversary of the invention of sound recording for reproduction and while audio technology has evolved significantly since the days of Edison's cylinder-based phonograph, the basic principles behind the act of recording have remained somewhat unchanged. To understand what makes us want to record sound, one must simply try to recall a unique moment that cannot be relived or a profound experience that cannot be replicated. Such memories make evident the natural human impulse to preserve anything that is evocative and singular. As an example, it is not uncommon for music fans to go to great lengths to record the live performances of their favourite artists in order to revisit them at a later date or to share them with absent friends.

Recording is mostly a social, collaborative activity and in this capacity it appears as more primal or instinctive than the commonly more isolated and lonelier processes of mixing and mastering.

As far as the preserving of one's own music, a passage in Wim Wender's film *Until the End of the World* may resonate with artists that record themselves frequently. In the aforementioned film, a scientist accidentally develops a machine that records dreams and its use becomes addictive to some of those who try it. Such individuals eventually appear to know themselves and each other more profoundly through the exchange of their 'recordings'.

The sound recorder is the 'mirror with a memory' for those who listen. With it we can capture glimpses of our reality, producing 'pictures' as we express ourselves. These may ultimately serve as reminders of who we were and of our dreams.

MUSIC PRODUCTION

RECORDING

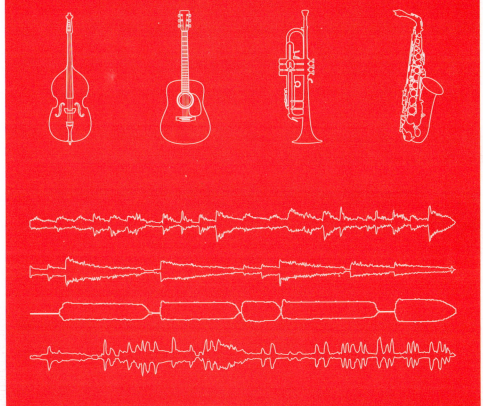

CORE CONCEPTS

RECORDING ENVIRONMENTS

Music is created in a multitude of environments, ranging from sophisticated and commonly expensive high-end recording studios to small bedrooms. With this in mind, serious recordists must be prepared to capture performances in any space considered suitable by music makers, including venues selected for reasons other than their sonic attributes (which may seem frustrating to those expecting to always operate under ideal circumstances).

The evaluation of a potential recording space is not simple, as a number of indirectly related and at times unpredictable factors may affect the result of sessions. In fact, in some circumstances the selection of recording settings based on emotional criteria may yield results that justify a sacrifice in sound quality, e.g. when comfort outweighs unfavourable acoustics.

Regardless of what supports the choice of location, recordists must be able to appraise and optimise their working conditions, ensuring that projects can be completed successfully within their given circumstances.

TRADITIONAL RECORDING STUDIOS

What is considered here as a traditional recording studio is any complex consisting of 'live' and control rooms, based around medium or large-format mixing consoles and offering all the other common elements of audio signal chains, e.g. cabling, microphones, signal processing devices, recorders, etc. The number of such studios has declined significantly since the introduction of affordable digital audio workstations and the proliferation of home studios, although many producers and artists still feel that they are the ideal place for music recording.

The following are a few advantages of working in well-established recording studios:

1. History / Tradition
Musicians commonly gain confidence when working in environments where other successful albums were made, e.g. a band recording at Abbey Road Studios for the first time will most likely confirm the power of tradition.

2. Equipment
The equipment found in established studios is commonly of high quality and it undergoes maintenance checks and repairs at regular intervals.

3. Acoustics / Environment

There is something to be said for working in an environment entirely dedicated to the making of music. The acoustics of high-end studios are the result of careful planning and significant investment.

4. Personnel

Studio personnel are required to know their premises in great detail in order to work efficiently and support visiting producers, engineers, musicians, etc.

5. Focus

High-end recording studio time is commonly expensive, which may lead producers, engineers and artists to focus and work with greater intensity. It is important to note that this may easily have the opposite effect on some individuals.

TEMPORARY / 'RESIDENTIAL' RECORDING SPACES

Residential recording spaces come in different shapes and sizes. Some share the characteristics of large commercial studios, while others may be substantially less sophisticated. The main advantages of recording in live-in settings arise from the possible isolation from the 'outside world'.

It is not uncommon for a collaborative bond to be created between production teams working in residential settings that would be difficult, if not impossible, to recreate in traditional studio environments.

PROJECT / HOME STUDIOS

'Project' and home studios have grown significantly in number since the introduction of affordable DAWs and today a great number of musicians own and are able to operate simple, computer-based, recording set-ups.

The following are some of the advantages of recording in a domestic environment:

- Artists are able to work whenever they feel inspired.
- Artists may feel less pressured to produce results.
- Introverted artists may feel more confident working in the comfort of their own home.
- Artists are able to save money that would otherwise be used for studio rental.

It is nevertheless important for home recordists to be aware of their set-up's limitations, where the quality and condition of equipment and the overall acoustic characteristics of the recording and monitoring environments may not allow for professional-sounding results to be achieved.

LOCATION / LIVE PERFORMANCE VENUES

What is considered here as work on location is any production carried out in environments that are not fully dedicated to the recording of music and that may only be accessed for a restricted length of time, i.e. settings that require the recording team to set up and break down immediately prior to, and after work respectively. The recording of live performances is a variant of such work, with the added element of 'pressure' originating from:

- The presence of a paying audience
- The possibility of 'live' broadcasting
- The likelihood of two independent teams working in tandem.

Although the recording of concerts is not the primary focus of this book, all the information included here may be applied to 'live' work.

THE RECORDING TEAM

A recording team may have some or all of the following elements:

- Musicians
- Programrs
- Runners
- Tape operator
- DAW operator / Editor
- Assistant engineer
- Main engineer
- Producer
- Others.

It is not always easy to delineate the roles and responsibilities of those involved in music production. This is particularly true in the case of modest budget projects, where individuals are commonly required to operate in multiple capacities and the labelling of roles and responsibilities may seem pointless. Still, in traditional recording environments a clear outline of functions often leads to a better distribution of labour and the more efficient use of time.

The following is a brief description of the members of conventional recording teams and their roles:

MUSICIANS

Musicians are defined here as the individuals that generate sound during the recording stage of a production. This applies to those that utilise both traditional and non-traditional instruments, e.g. turntables, etc.

PROGRAMMERS

Programrs may contribute to music production by selecting or creating electronic instrument 'patches', programming drumbeats, generating sequences, etc., which may occur during the pre-production, recording or the mixing stages.

RUNNERS

Runners (or 'tea boys') are commonly the youngest members of the recording team. Their tasks can be varied and commonly include peripheral or indirectly related activities that help sessions flow smoothly, e.g. purchasing of media, catering, etc.

THE TAPE OPERATOR

The 'tape op' is responsible for the basic maintenance and the operation of tape recorders. This position is no longer common in music production.

THE DAW OPERATOR / EDITOR

A DAW operator should know audio software programs and their corresponding hardware in great detail, being able to work with them during the recording, editing and mixing stages of production.

THE ASSISTANT ENGINEER

This role requires the most flexibility, as the duties of the assistant engineer may span from those of a 'runner' to those of a main recording engineer. It is presently not uncommon for assistant engineers to also be responsible for tasks that were normally assigned to tape or DAW operators.

THE MAIN ENGINEER

The main recording engineer is the person who is ultimately accountable for the successful recording of audio onto the chosen medium, although this does not imply he or she is responsible for the aesthetic-related attributes of the recorded material.

THE PRODUCER

This role varies according to context. In classical music, the producer is commonly the decision-maker regarding performance while in pop / rock he or she traditionally acts as a general manager, being ultimately responsible for the successful completion of a project. It is also important to note that in electronic / dance circles, composers / programrs are frequently referred to as 'producers'.

OTHERS
- Composer / Arranger

- A&R / Record company representatives
- Artist managers
- Investors.

A number of other professionals may be indirectly involved in the recording stage of a project. Some of them may appear somewhat disconnected from the artistic process, although anyone involved in production should be aware that investors of time and / or money are likely to expect the right to an opinion on the product they help generate.

THE TOOLS

Advanced technicians from different fields seem to share the ability to select and use their tools skilfully, achieving consistent high-quality results. Such proficiency commonly results from years of experience, although it may be mistaken for a natural talent. In fact, technicians that develop their expertise through practice may find it difficult to rationalise or explain their actions, as production appears progressively more intuitive and effortless.

Although technical knowledge alone does not guarantee successful results, it may provide the necessary foundation upon which experience may be built. With this in mind, it seems advisable for recordists to study the tools they work with in detail, setting out guidelines for their selection and use. To some, this process could initially be based on familiarity or external influences, e.g. using Neumann microphones to record vocals, as Geoff Emerick did to record the Beatles, although ultimately it would be more appropriate for such important decisions to be based on a balance between intuition, scientific knowledge, experience and the desire to experiment.

The following sections present possible strategies for the selection of recording tools.

MICROPHONES

Microphones are broadly described as transducers or converters of energy. These, alongside loudspeakers, arguably constitute the most important elements in the audio chain, as both perform vital roles at the two ends of the signal path. While the microphone selection process may seem simple and spontaneous to some, it is not uncommon for recording engineers to spend considerable time and effort to ensure the best device is selected for each given job.

While the criteria influencing the choice of transducers may extend well beyond the realm of technical specifications, for the sake of objectivity four main factors are presented here as most important for microphone selection. These are:

- Transducer type – Bias vs. Fidelity
- Diaphragm size – Output sensitivity / Bias vs. Fidelity
- Directionality – Separation
- Construction – Physical considerations.

MICROPHONE SELECTION CRITERION 1: TRANSDUCER TYPE (BIAS vs. FIDELITY)

A basic understanding of the physical principles of transduction on the part of the reader is expected in this section, where the main objective is to discuss the general characteristics and merits of different transducer types, which are described next.

Transducers may be classified according to their operating principle, falling into the following key groups:

- Electro-dynamic or dynamic (moving coil, ribbon and printed ribbon)
- Condenser (Including electret condenser and PZM or boundary)
- Piezo or crystal-based
- Other.

Moving Coil-Based Dynamic Microphones

Moving coil microphones:

- Have relatively heavy moving parts and therefore present slower transient response and lower sensitivity to small variations in pressure (less detailed pickup).
- Have limited high-frequency response.
- Are commonly not 'transparent', i.e. they generally add 'colour' in the high-mid frequency range (~ 5 kHz to 10 kHz).
- Are robust and can handle high sound pressure levels.
- Are effective when placed in close proximity to the sound source.

Ribbon-Based Dynamic Microphones

Ribbon microphones:

- Have light moving parts and are sensitive to small variations in pressure, commonly presenting a more accurate transient response than moving coil microphones.
- Usually add 'colour' to recordings (roll-off of high frequencies).
- May sound 'thin' if placed distant from sound sources.
- Present a pronounced bass response when placed close to performers ('proximity effect').
- Are normally more fragile than moving coil microphones.

Printed Ribbon-Based (Regulated Phase) Microphones

In the less popular 'printed ribbon' or regulated phase transducer, an aluminium spiral ribbon is mounted onto a plastic (mylar) diaphragm, which is in turn placed between two ring magnets. Regulated phase transducers:

- Have light moving parts (good transient response).
- Tend to be more resilient than traditional ribbon microphones.

The following illustrations depict examples of moving coil and ribbon-based dynamic microphones that are commonly used in music production. The regulated phase example is included here for historical / illustrative purposes, as this type of transducer is not currently popular and its use has become somewhat rare.

DYNAMIC
MOVING COIL MICROPHONES

SHURE BETA 57A

Transducer: Moving coil
Pattern: Supercardioid
Response: 50 Hz - 16 kHz
Sensitivity: 2.7 mV/Pa
Max. SPL: ~140 dBSPL

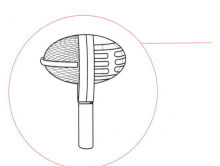

AKG D112

Transducer: Moving coil
Pattern: Cardioid
Response: 20 Hz - 17 kHz
Sensitivity: 1.8 mV/Pa
Max. SPL: >160 dBSPL

DYNAMIC
MOVING COIL MICROPHONES

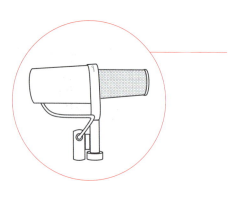

SHURE SM7

Transducer: Moving coil
Polar Pattern: Cardioid
Response: 40 Hz - 16 kHz
Sensitivity: 1.12 mV/Pa
Max. SPL: ~140 dBSPL

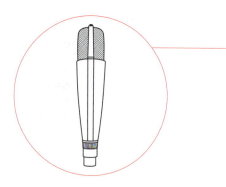

SENNHEISER MD421

Transducer: Moving coil
Polar Pattern: Cardioid
Response: 30 Hz - 17 kHz
Sensitivity: 2 mV/Pa
Max. SPL: N/A

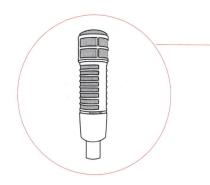

ELECTROVOICE RE20

Transducer: Moving coil
Polar Pattern: Cardioid
Response: 45 Hz - 18 kHz
Sensitivity: 2.8 mV/Pa
Max. SPL: N/A

DYNAMIC
RIBBON MICROPHONES

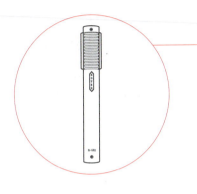

ROYER R121

Transducer: Ribbon
Pattern: Figure-of-8
Response: 30 Hz - 15 kHz
Sensitivity: 2 mV/Pa
Max. SPL: ~135 dBSPL

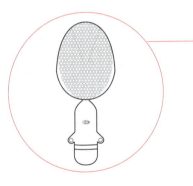

COLES 4038

Transducer: Ribbon
Pattern: Figure-of-8
Response: 30 Hz - 15 kHz
Sensitivity: 0.5 mV/Pa
Max. SPL: ~125 dBSPL

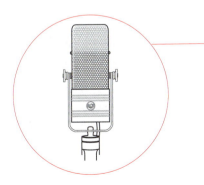

RCA 44BX

Transducer: Ribbon
Pattern: Figure-of-8
Response: 30 Hz - 16 kHz
Sensitivity: N/A
Max. SPL: N/A

DYNAMIC PRINTED RIBBON MICROPHONE

FOSTEX M11RP

Transducer: Printed Ribbon
Pattern: Cardioid
Response: 40 Hz - 18 kHz
Sensitivity: 2.8 mV/Pa
Max. SPL: ~130 dB$_{SPL}$

Condenser Microphones

Condenser microphones rely on the varying distance between the two plates of a capacitor to operate (variable capacitance).

Condenser-based transducers:
- Have very light moving parts, which allow for a fast transient response and high sensitivity.
- Offer an extended high-frequency response.
- Can be extremely transparent (flat frequency response).
- Commonly require a (phantom) power supply to operate.
- Frequently incorporate a built-in amplifier section (valve or FET), used to boost output gain and convert (lower) impedance.
- May be placed at greater distances from sound sources.
- Are commonly fragile.

Electret Condenser Microphones

The difference between electret and 'true' condenser microphones lies in the method used for diaphragm polarisation. In broad practical terms, the permanently polarised backplate of 'electrets' makes it possible to construct of a less expensive and more resilient microphone that may be placed in extreme proximity to sound sources, e.g. miniature instrument clip-on condenser microphones.

Pressure Zone Microphones

Pressure zone microphones (PZM), also known as 'boundary microphones' were developed in an attempt to avoid phase interference caused by hard surface reflections (in close proximity to the sound source).

In pressure zone microphones a miniature condenser capsule is found at a very short distance from a metal boundary, which is commonly placed on reflective surfaces, such as walls, piano lids, etc.

Boundary microphones:
- Do not present off-axis colouration.
- Present an even frequency response and high sensitivity.
- Commonly offer good signal to noise ratios.

Valve (Tube) Microphones

The built-in amplifier found in condenser microphones may rely on valve or transistor-based circuitry to operate. Valve-based condensers typically require a dedicated power supply unit, capable of delivering voltages in the 100 V range and are usually connected to their power supplies via multi-pin (five to seven-pin) cables.

It is advisable for recordists to power their valve-based (tube) condenser microphones at least one hour before sessions, as such devices commonly require time to reach a consistent, stable level of operation.

Digital Microphones

Digital microphones are commonly comprised of a regular condenser capsule followed by an analogue to digital converter. This allows for the conversion of audio at an early stage, which helps minimise the degradation of signals. Some digital microphones incorporate extra features such as peak limiting and remote control over polar pattern selection, input pad, pass-filter, etc.

Phantom Power

Some microphones require a supply of a DC voltage to operate. This supply is commonly delivered at 48 volts DC via XLR pins two and three and it is used to power the built-in amplifier used for impedance matching (following the transducer). This improves signal transmission over longer cable runs. In the case of non-electret (true) condenser microphones, phantom power is also used to polarise the backplate of the capsule.

Recordists should be very careful when using phantom power, where the following rules must be observed:

- Always mute the control room monitors before switching phantom power on or off.
- Avoid sending phantom power to microphones that do not require a DC supply, as these may get damaged, e.g. through cross-patching (patchbays).
- Be aware of the danger of using gender changer cables and avoid the presence of phantom power in male XLR connectors (due to the risk of electric shock).
- Patch all XLR cables onto their respective microphones prior to switching phantom power on.
- Turn phantom power off before disconnecting all condenser microphones.

Always neutralise mixing consoles after sessions, ensuring phantom power is off in all channels.

The following pages contain examples of condenser microphones that are commonly used in music production.

CONDENSER MICROPHONES

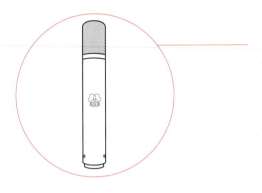

AKG C12

Transducer: Condenser
Pattern: Variable
Response: 30 Hz - 15 kHz
Sensitivity: ~10 mV/Pa
Max. SPL: ~115 dBSPL

NEUMANN U47

Transducer: Condenser
Pattern: Cardioid
Response: 30 Hz - 15 kHz
Sensitivity: ~25 mV/Pa
Max. SPL: ~110 dBSPL

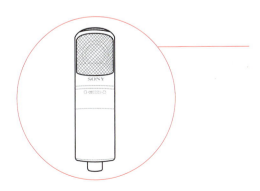

SONY C800G PAC

Transducer: Condenser
Pattern: Omni / Cardioid
Response: 20 Hz - 18 kHz
Sensitivity: ~25 /Pa (omni) /
18 mV/Pa (cardioid)
Max. SPL: ~113 dBSPL

CONDENSER MICROPHONES

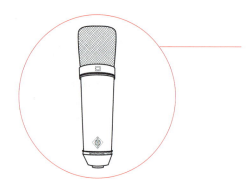

NEUMANN U87

Transducer: Condenser
Pattern: Variable
Response: 20 Hz - 20 kHz
Sensitivity: 28 mV/Pa
Max. SPL: 117 dBSPL

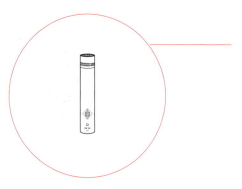

NEUMANN KM184

Transducer: Condenser
Pattern: Cardioid
Response: 20 Hz - 20 kHz
Sensitivity: 15 mV/Pa
Max. SPL: 138 dBSPL

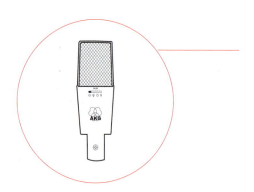

AKG C414B-ULS

Transducer: Condenser
Pattern: Variable
Response: 20 Hz - 20 kHz
Sensitivity: 12.5 mV/Pa
Max. SPL: 134 dBSPL

CONDENSER MICROPHONES

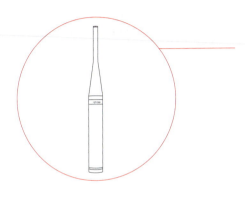

EARTHWORKS QTC50

Transducer: Condenser
Pattern: Omni
Response: 3 Hz - 50 kHz
Sensitivity: 30 mV/Pa
Max. SPL: 142 dB$_{SPL}$

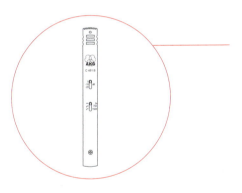

AKG C451B

Transducer: Condenser
Pattern: Cardioid
Response: 20 Hz - 20 kHz
Sensitivity: 9 mV/Pa
Max. SPL: 155 dB$_{SPL}$

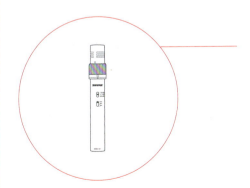

SHURE KSM141

Transducer: Condenser
Pattern: Variable
Response: 20 Hz - 20 kHz
Sensitivity: 14 mV/Pa
Max. SPL: 134 dB$_{SPL}$

PZM MICROPHONES

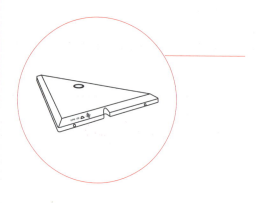

NEUMANN GFM132

Transducer: Condenser
Polar Pattern: Hemispherical
Response: 20 Hz - 20 kHz
Sensitivity: 18 mV/Pa
Max. SPL: 137 dBSPL

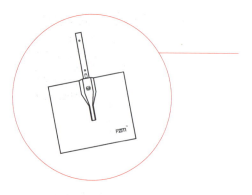

CROWN PZM-30D

Transducer: Condenser
Polar Pattern: Hemispherical
Response: 20 Hz - 20 kHz
Sensitivity: 7 mV/Pa
Max. SPL: 150 dBSPL

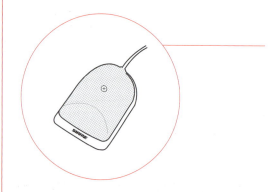

SHURE EZBO

Transducer: Condenser
Polar Pattern: Hemispherical
Response: 50 Hz - 17 kHz
Sensitivity: 11 mV/Pa
Max. SPL: 118 dBSPL

Piezo Microphones

Piezo or crystal-based capsules are commonly employed in 'contact' microphones or acoustic instrument 'pick-ups'.

Piezo microphones:
- Are commonly biased towards high / high-mid pick up, with poor low and very high-frequency representation.
- Minimize the effect of feedback originating from backline.
- Work well to capture signals for subsequent replacement, e.g. snare drum contact microphones.

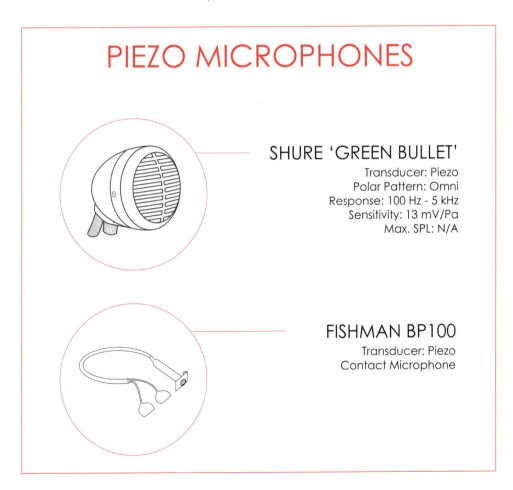

PIEZO MICROPHONES

SHURE 'GREEN BULLET'
Transducer: Piezo
Polar Pattern: Omni
Response: 100 Hz - 5 kHz
Sensitivity: 13 mV/Pa
Max. SPL: N/A

FISHMAN BP100
Transducer: Piezo
Contact Microphone

OTHER MICROPHONES

Other types of microphone exist, e.g. carbon, fibre-optic, laser, etc. although these are not commonly used in music applications at present.

MICROPHONE SELECTION CRITERION 2: DIAPHRAGM SIZE
(OUTPUT SENSITIVITY / FIDELITY)

The influence of diaphragm size over microphone performance varies according to transducer type. Most authors question the use of large capsules in dynamic microphones, arguing that due to operating principle the benefits of such implementation are minimal, if not negligible. A few designers, such as Bob Heil, believe on the other hand that a larger diaphragm equates to improved transient response, even in the case of moving coil microphones, as long as the parts are made light enough. Regardless of the debate, capsule dimensions have a more tangible effect in the case of condenser microphones, where:

1. Microphone sensitivity is commonly proportional to diaphragm size.
2. Off-axis pickup fidelity is inversely proportional to diaphragm size.

As a general rule, the use of a larger sized diaphragm should be considered when significantly small variations in pressure must be captured with great detail, e.g. vocal performances where minute 'mouth' noises may add intimacy to a recording. At the same time, the use of small diaphragm transducers is indicated when off-axis sounds must be captured with great fidelity, e.g. large ensembles recorded with few microphones.

It is important to note that sound engineering authors do not seem to agree on the impact of diaphragm size over bandwidth. Considering the series of factors that may influence a microphone's response, e.g. diaphragm tension, mass density and the sizes of the internal cavity and of the static pressure equalization vent, it does not seem reasonable to assume there is a simple causal relationship between diaphragm size and bandwidth of pickup, i.e. recordists should always check a microphone's frequency response graph.

MICROPHONE SELECTION CRITERION 3: DIRECTIONALITY (SEPARATION)

Two important questions provide the basis for this criterion:
1. How much separation between channels is required?
2. How much natural reverberation is desired?

If in answering the first question it is established that a great degree of separation is required, omnidirectional or bi-directional microphones should be avoided or used with caution (unless the physical isolation between performers in the studio is achievable and practical). In such case, directional microphones, e.g. hypo, super, hyper or standard cardioid transducers represent a better choice, as these may offer considerable rejection of unwanted sounds. This is particularly important in the case of projects where extensive editing may be required, i.e. productions in which song arrangement or structure might change and different instrumental parts may need to be moved or copied between sections.

As far as the second question is concerned, transducers with wider pick-up patterns, e.g. omni and hypocardioid, may capture the natural sound of an environment with more accuracy. In favourable recording settings, the use of non-directional microphones may prove invaluable and should be explored.

It is important to mention that directionality will also affect frequency response (and therefore 'fidelity') for both on and off-axis sound pick-up.

Polar Patterns

Polar pattern graphs may display one (1 kHz) or many curves, e.g. 100 Hz, 10 kHz, etc., as microphones do not present a constant pick-up pattern across the audio spectrum, i.e. polar patterns change according to frequency and microphones become less directional when handling material at the extremes of range.

1. Omnidirectional microphones:
   * Are equally sensitive to sound arriving from all directions.
   * Provide the flattest frequency response with no proximity effect (if not based on dual cardioid capsules combined).
   * Are less prone to the pick up of rumble and wind sounds.
   * Present the least coloration of off-axis sounds.

2. Bi-Directional / microphones:
   * Are most sensitive to sounds arriving from two directions, i.e. the front and the back of the diaphragm.
   * Exhibit a prominent 'proximity effect'.

3. Unidirectional microphones (Hypo / Super / Hyper / Cardioid):
   * Are most sensitive to sounds arriving from one direction (on axis).
   * Exhibit the 'proximity effect'.

The following pages describe common microphone polar patterns, including a brief examination of their use.

POLAR PATTERNS

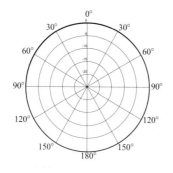

Omnidirectional

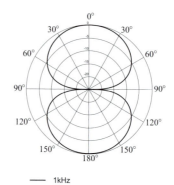

Figure-of-Eight

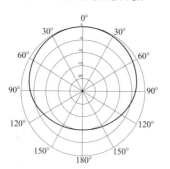

Hypocardioid

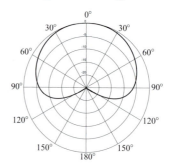

Cardioid

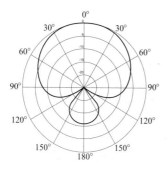

Supercardioid

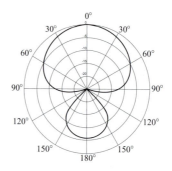

Hypercardioid

EXPLORING POLAR PATTERNS

OMNIDIRECTIONAL

Particularly suitable for the recording of:

- Background vocals
- Large ensembles, e.g. orchestras
- Instrument sections, e.g. horns and strings
- Room ambiance, e.g. mono overheads
- 'Vintage' sounding, period music
- Mono samples.

In all the aforementioned examples, the sound source(s) may be positioned up to ± 180° (360°) around the transducer.

HYPOCARDIOID

Similar in application to omnidirectional capsules, although better suited for the rejection of undesired off-axis sounds, such as:

- Audience noise (live performances)
- Excessive reverberation originating from walls opposite to the performers.

Sound sources should ideally be positioned in front of the transducer (at up to approx. ± 60°).

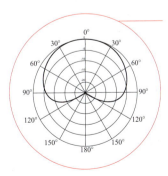

CARDIOID

Very efficient when used to record:

- Musicians facing one another (where separation is desired)
- Sources in venues that are excessively reverberant / noisy
- Guitar / Bass amplifiers, when these must be recorded in the same room as other instruments, e.g. drums (amps placed opposite to the drum kit).

EXPLORING POLAR PATTERNS

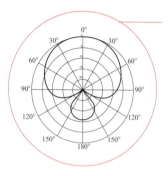

SUPERCARDIOID

Useful when sounds originating from the sides of the transducers must be rejected e.g. portions of a large ensemble (commonly used as 'spot' microphones).

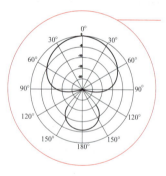

HYPERCARDIOID

Similar to supercardioid, while offering extended rejection of sounds originating from the immediate sides of the transducer e.g. useful for avoiding hi-hat 'spill' on snare drum recordings.

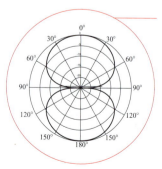

FIGURE-OF-8

Suitable for the simultaneous recording of musicians e.g. singers, placed opposite each other (when separation is **not** desired).

Useful for the recording of singing acoustic guitarists, where two transducers are placed above one another, pointing at the performer's mouth and at the guitar (rejection at ± 90°).

Proximity Effect

The so-called 'proximity effect' may be described as a directional microphone's increase in bass response as the sound source approaches the diaphragm.   This effect is commonly explored artistically, where the signal from 'thin' sounding performers is enhanced through close microphone placement.

MICROPHONE SELECTION CRITERION 4: CONSTRUCTION (PHYSICAL CONSIDERATIONS)

This last criterion is the simplest, where selection is influenced or dictated by the overall physical dimension of the microphone, e.g. when difficult placement or restricted access may require the use of a transducer of small size. In such circumstances a compromise may be required, where sound quality may not be given priority over other criteria.

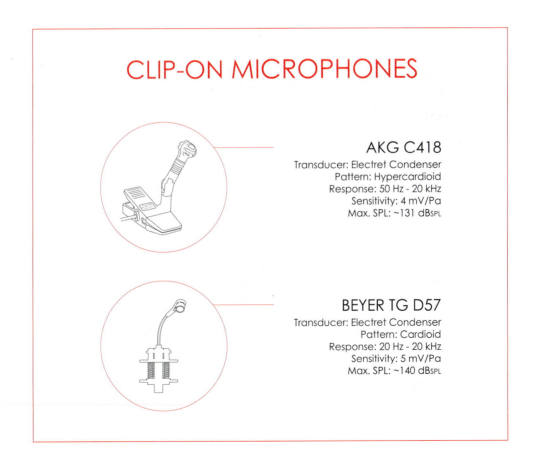

CLIP-ON MICROPHONES

AKG C418
Transducer: Electret Condenser
Pattern: Hypercardioid
Response: 50 Hz - 20 kHz
Sensitivity: 4 mV/Pa
Max. SPL: ~131 dBSPL

BEYER TG D57
Transducer: Electret Condenser
Pattern: Cardioid
Response: 20 Hz - 20 kHz
Sensitivity: 5 mV/Pa
Max. SPL: ~140 dBSPL

MOVING COIL vs. CONDENSER SELECTION CHART

The following guidelines apply to the selection of microphones containing the most common types of transducers: the moving coil and the condenser-based.

A. High output (sensitivity) and/or
B. Extended (upper) frequency range pick-up,
C. Flat frequency response,
D. Excellent response to quiet transients required?

A. High-impact / amplitude area of placement
B. Biased ('flattering') pick-up
C. 'Softening' of extreme transients desired?

Condenser transducer

Moving-coil dynamic transducer

Potentially very quiet, highly detailed, (mostly) 'on-axis' sound source

Wide sound source, possibly surrounding the transducer

Enhanced rejection of off-axis sound desired?

Large Diaphragm

Small Diaphragm

YES   NO

Isolation / Separation and/or High direct to reverberant sound ratio desired?

Isolation / Separation and/or High direct to reverberant sound ratio desired?

Supercardioid or Hypercardioid pattern

Cardioid pattern

YES   NO

YES   NO

Unidirectional pattern

Proximity effect desired?

Unidirectional pattern

Proximity effect desired?

YES   NO

YES   NO

Figure-of-8 pattern

Omni pattern

Figure-of-8 pattern

Omni pattern (flattest frequency response)

DI / DIRECT INJECTION BOXES

A DI or direct injection box is a device that allows for high impedance electric or electronic instruments to be connected to low impedance inputs, e.g. through direct injection, the output of an electric bass or guitar may be fed to a console's microphone input, bypassing the use of an instrument amplifier (backline) and of a microphone.

Passive DI boxes lower the impedance of the source (the instrument) through the use of transformers. The latter also allow for the balancing of signals, i.e. a conversion from unbalanced 'jack' to balanced XLR connections. Active DI boxes utilise electronic circuitry to achieve the same impedance conversion as passive devices, while offering improved fidelity, i.e. higher gain and wider frequency response. Such devices are commonly fed by phantom power or by batteries.

DI boxes make it possible for:
- High-impedance electric or electronic instruments to be recorded without backline amplification.
- Instruments to be recorded in extreme isolation.
- Signals to be transmitted over longer distances (through balancing).
- The subsequent 're-amping' of instrument signals.

It is important to point out that active direct injection boxes may not be perceived as 'superior' to passive devices by many musicians who prefer using the latter.

Parallel Outputs
It is common for DI boxes to offer a 'jack' connector parallel output patch point on their 'input' side, allowing for the simultaneous feed of a console and of an instrument amplifier. In such cases it is highly recommended for engineers to record both signals, as this may allow for the combining of the two tracks during mixdown.

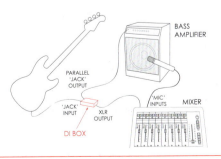

DI BOXES

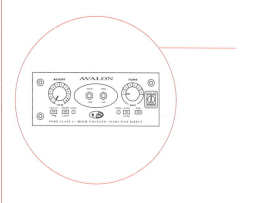

AVALON ULTRA FIVE
Active DI Box

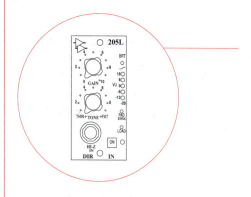

API 205L
Active DI Box

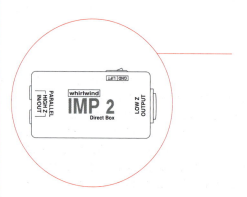

Whirlwind IMP 2
Passive DI Box

Impedance

Impedance is total opposition to current flow, i.e. the combination of resistance (frequency independent) and reactance (frequency dependent) opposition. Output impedance may be broadly described as a measure of how difficult it is to 'extract' signals from an electric or electronic sound source.

Ideally, a lower output impedance should meet a higher input impedance, e.g. microphones with impedance ranging from 50 to 600 ohms should normally feed preamplifiers with impedances ranging from 2000 to 5000 ohms. This helps ensure that signals will reach the input or 'load' with fidelity. DI boxes lower the high output impedance of electric instruments so they may feed microphone preamplifers directly (a 10^4 to 10^2 range conversion).

Balancing

Balancing is an elegant solution for the problem of electromagnetic interference. In balanced connections, an original audio signal ('hot') and a polarity-inverted copy ('cold') travel along a twisted pair of conductors. In the case of electromagnetic interference, both conductors are equally affected and the rejection of unwanted sounds is made possible through the polarity inversion of the 'cold' component at the destination (or input).

It is important to note that in the case of short cable runs, the use of balanced connections may not be necessary or justifiable.

Ground 'Lifting'

DI boxes commonly offer a 'ground lift' option. This can be useful when an instrument, e.g. a bass, is connected simultaneously to a mixing console and to an amplifier (backline) at different potentials to the ground. Such arrangement can lead to an AC current flowing in the shield of the conductor, which may ultimately lead to 50 or 60 Hz cycle 'hum'. The ground lift function decouples the input and the output of the circuit electronically, commonly reducing or eliminating 'hum'.

Ground 'Lifting' (continued)
DI boxes should be set originally on their 'ground' position and only set onto 'lift' in the case of a ground loop problem.

Valve (Tube) DI Boxes
Valve (tube) DI boxes utilise electronic circuitry to lower impedance and to amplify and balance signals.

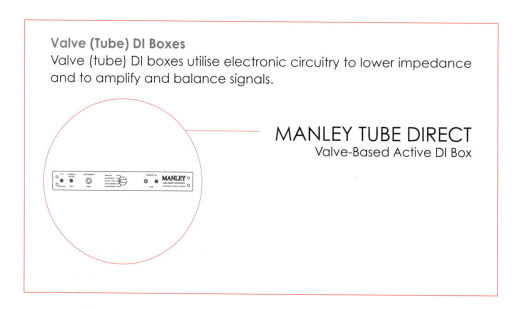

MANLEY TUBE DIRECT
Valve-Based Active DI Box

ANALOGUE AUDIO CABLES / CONNECTORS

'Jack' / Phone
The following are types of 'jack' or phone connectors:
- 3.5 mm / 1/8" Jack or 'TS' 'Minijack'
  Two terminals: tip and sleeve
- 3.5 mm / 1/8" Jack or 'TRS' 'Minijack'
  Three terminals: tip, ring and sleeve
- Bantam / TT
  Three terminals: tip, ring and sleeve – 'TRS'
- BPO (BPO316) / 'B-gauge'
  Three terminals: tip, ring and sleeve – 'TRS'
- 6.35 mm / 1/4" Jack 'TS' 'A-gauge'
  Two terminals: tip and sleeve
- 6.35 mm / 1/4" Jack 'TRS' 'A-gauge'
  Three terminals: tip, ring and sleeve
- 2.5 mm Jack (not commonly used in current manufacturing).

'JACK' CONNECTORS

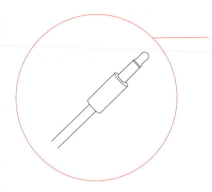

3.5 mm / 1/8" 'TS' MINIJACK
Semi-Pro / Amateur audio equipment

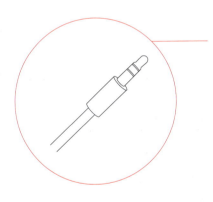

3.5 mm / 1/8" 'TRS' MINIJACK
Headphones
MP3 Players
Computers

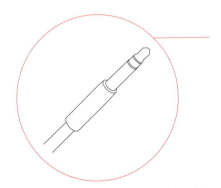

BANTAM (TT)
Professional patchbays

'JACK' CONNECTORS

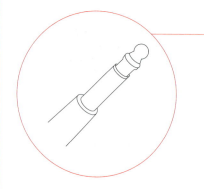

BPO
Professional patchbays

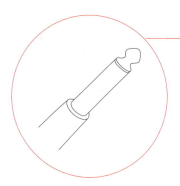

6.35 mm / 1/4" 'TS' JACK
Unbalanced electric / electronic instruments
Unbalanced console connections
Pro/Semi-pro patchbays
Effects processors

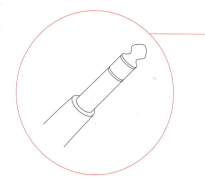

6.35 mm / 1/4" 'TRS' JACK
Balanced electronic instruments
Balanced console connections
Pro/Semi-pro patchbays
Effects processors
Audio Interfaces
Tape Recorders
Headphones

RCA / Phono

RCA or 'phono' connectors commonly contain two terminals (tip and sleeve – 'TS').

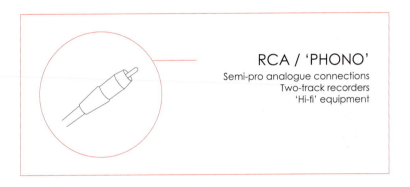

RCA / 'PHONO'
Semi-pro analogue connections
Two-track recorders
'Hi-fi' equipment

XLR / Cannon

XLR or Cannon connectors / cables are available in two, three, five and seven-pin configurations.

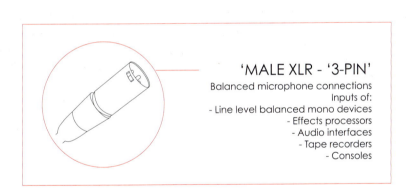

'MALE XLR - '3-PIN'
Balanced microphone connections
Inputs of:
- Line level balanced mono devices
- Effects processors
- Audio interfaces
- Tape recorders
- Consoles

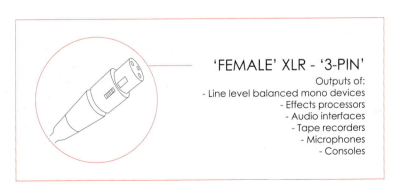

'FEMALE' XLR - '3-PIN'
Outputs of:
- Line level balanced mono devices
- Effects processors
- Audio interfaces
- Tape recorders
- Microphones
- Consoles

5-Pin XLR Connectors

5-pin XLR connectors are employed by some dual-capsule stereo microphones.

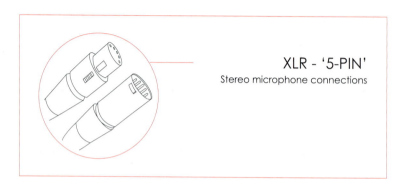

XLR - '5-PIN'
Stereo microphone connections

7-Pin XLR Connectors

7-pin XLR connectors are used for the connection of some valve microphones to their corresponding power supplies.

Speakon

There are three common types of speakon connectors:
NL2 – containing two terminals.
NL4 – containing four terminals.
NL8 – containing eight terminals.

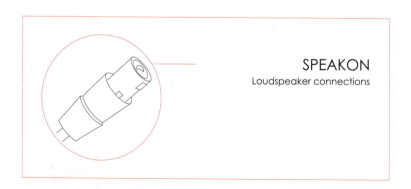

SPEAKON
Loudspeaker connections

'Banana'

'Banana' connectors contain one terminal.

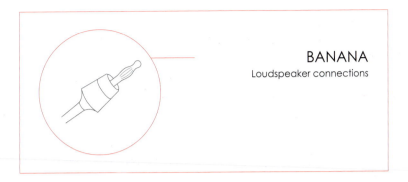

BANANA
Loudspeaker connections

D-Subminiature / D-Sub

There are several types of multi-pin D-subminiature connectors, e.g. DE9, DA15, DB25, DE26, DC37, DD50. The most common type of audio D-sub connector is the DB25, which contains 25 terminals.

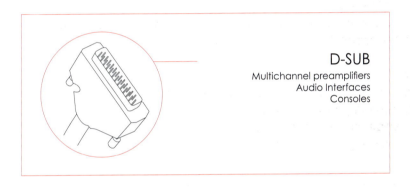

D-SUB
Multichannel preamplifiers
Audio Interfaces
Consoles

EDAC / ELCO

EDAC or ELCO connectors are available in 20, 38, 56, 90 and 120-pin configurations.

EDAC
Multichannel preamplifiers
Audio Interfaces
Tape recorders
Consoles

Shielding

Metallic shielding (foil or braided) is commonly employed in audio cable manufacturing in an attempt to minimise the effects of noise. Shielded cables may be coaxial, triaxial or quadraxial depending on their conductor/shield configuration.

Cable Runs

Recordists must aim to use audio cables of suitable short length, laying them away from current carrying conductors, i.e. power cables, in order to:

- Minimise possible electromagnetic interference (especially on unbalanced lines).
- Avoid high-frequency roll-off (due to capacitance, where a long cable may act as a filter).

Any unavoidable extra cable length ('slack') must be laid on a 'figure-of-8' pattern for the reduction of the effects of electromagnetic interference, as improperly wound cables may act as 'aerials' and/or high-cut filters.

If an overlap between audio and electrical cables is unavoidable, these should be set at a 90° angle, so the effects of interference / induction are minimised.

It is important for recordists to be able to assemble audio cables by soldering conductors to their respective connector terminals. Soldering is an invaluable skill, particularly for those working in isolated residential music production environments. The following pages describe the conductor / connector configuration for audio cables that are commonly used in music production.

NB All balanced output to unbalanced input connections depicted here correspond to impedance-balanced or transformer-based balanced devices feeding unbalanced inputs. In the case of active-based balanced devices feeding unbalanced inputs, the 'ring' of 'jack' connectors or 'pin-3' of XLR connectors should be left unconnected or 'floated'. All unbalanced feeds to balanced inputs should use 3 conductor cables and incorporate a 'cold' to 'ground' link on the output side only.

ANALOGUE CABLE WIRING

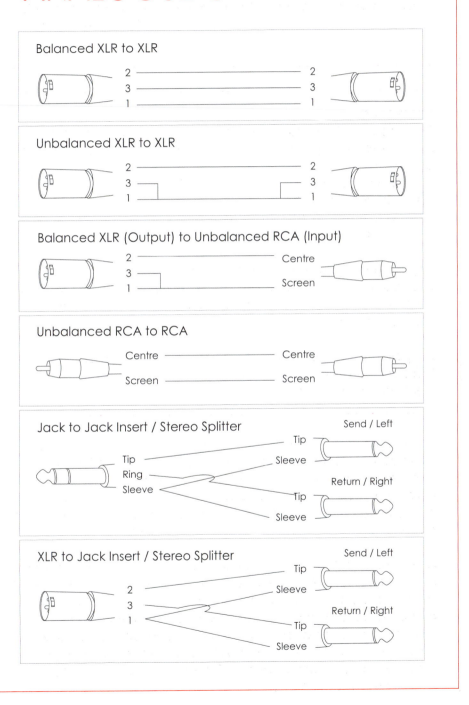

Balanced XLR to XLR

2 —————————— 2
3 —————————— 3
1 —————————— 1

Unbalanced XLR to XLR

2 —————————— 2
3 ——┐   ┌—— 3
1 ——┘   └—— 1

Balanced XLR (Output) to Unbalanced RCA (Input)

2 —————————— Centre
3 ——┐
1 ——┘   —————— Screen

Unbalanced RCA to RCA

Centre —————————— Centre
Screen —————————— Screen

Jack to Jack Insert / Stereo Splitter

Send / Left

Tip
Tip
Ring   Sleeve
Sleeve

Return / Right

Tip
Sleeve

XLR to Jack Insert / Stereo Splitter

Send / Left

Tip
2
3   Sleeve
1

Return / Right

Tip
Sleeve

ANALOGUE CABLE WIRING

Balanced Mono / Unbalanced Stereo Jack to Jack

Tip —————————— Tip
Ring —————————— Ring
Sleeve —————————— Sleeve

Unbalanced Stereo Jack to Jack

Tip —————————— Tip
Sleeve —————————— Sleeve

Balanced (Output) to Unbalanced (Input) Jack to Jack

Tip —————————— Tip
Ring
Sleeve —————————— Sleeve

Balanced Jack to XLR

Tip —————————— 2
Ring —————————— 3
Sleeve —————————— 1

Unbalanced Jack (Output) to Balanced XLR (Input)

Tip —————————— 2
3
Sleeve —————————— 1

Unbalanced Jack to RCA

Tip —————————— Centre
Sleeve —————————— Screen

Balanced Jack (Output) to Unbalanced RCA (Input)

Tip —————————— Centre
Ring
Sleeve —————————— Screen

DIGITAL AUDIO CABLES / CONNECTORS

TDIF (Tascam Digital Interface)
TDIF cables commonly terminate at 25-pin D-Sub connectors.

S/PDIF (Sony / Phillips Digital Interface)
S/PIDF protocol employs either coaxial low impedance (75 ohms) cables and RCA connectors, or fibre-optic cables and TOSLINK connectors.

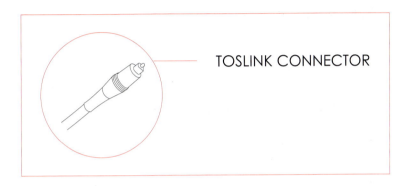

TOSLINK CONNECTOR

ADAT Lightpipe
ADAT Lightpipe utilises fibre-optic cables and TOSLINK connectors.

AES / EBU
AES/EBU digital cables comprise low impedance (110 ohms) conductors that commonly terminate at XLR 3-pin or D-Sub connectors.

MADI
The MADI protocol employs low impedance (75 ohms) conductors and BNC connectors or, less commonly, multimode fibre-optic cables.

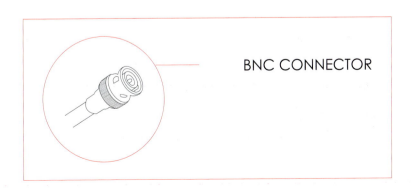

BNC CONNECTOR

OTHER CABLES / CONNECTORS

Word Clock
Word clock cables utilise low impedance (75 ohms) conductors and BNC connectors.

Word Clock Termination
Some digital device set-ups require word clock 'termination' in order to avoid cable reflection and possibly 'jitter'. Termination is achieved through the use of BNC 'T' connectors in conjunction with 75-ohm termination plugs. Users should check operation manuals to establish whether a chain requires external termination.

MIDI
MIDI cables utilise three conductors and 5-pin DIN connectors, hence only three of the five pins are wired (pins 2, 4 and 5). MIDI cables should run for a maximum of 15 metres or 50 feet approximately.

5-PIN DIN CONNECTOR

MIDI Through XLR / TRS Cables
Some recording sessions may require the transmission of MIDI signals between live and control rooms. In such cases, it may be necessary to use XLR or TRS 'jack' connections, as most wall boxes do not offer 5-pin connections. When wiring a 5-pin DIN to XLR or 'jack' lead, the three middle DIN pins (2, 4 and 5) should be linked to the three terminals of the XLR / 'jack' connector. The transmission of data should work as long as the configuration is kept consistent.

MIDI Through XLR / TRS Cables (continued)
The following is a possible way to wire a 5-pin DIN to XLR or TRS cable:
Pin 4 of 5-Pin DIN to pin 2 of XLR or 'tip' of TRS 'jack'
Pin 5 of 5-Pin DIN to pin 3 of XLR or 'ring' of TRS 'jack'
Pin 2 of 5-Pin DIN to pin 1 of XLR or 'sleeve' of TRS 'jack'

NB Such cables should not exceed approximately 2 metres or 6.5 feet in length.

MICROPHONE PREAMPLIFIERS

The main role of the microphone preamplifier is to raise microphone level signals to line level, although some units may offer extra features such as equalisation and compression. Microphone preamps differ in sound and while some devices may be broadly described as 'transparent', others alter the nature of audio material noticeably. In addition to that, equipment deemed as 'transparent' will commonly respond to quiet and loud sounds in a non-linear fashion, e.g. with increasing harmonic distortion.

A preamp's alteration of timbre must not necessarily be seen as a negative side effect of amplification, as experienced engineers frequently attempt to 'shape' microphone signals before routing them to the multitrack recorder, e.g. via the use of equalisation. With that in mind, it appears as a sensible procedure for recordists to audition a few different preamplification units and to do so at 'clean', i.e. conservative input signals, and at higher, 'overdriven' levels before committing to a signal chain.

Harmonic Distortion
Preamplifiers, as all other signal processors, add harmonic content to their input signal. The level and nature of the added harmonics or distortion is an important factor influencing the efficiency and the perceived quality of devices.

It is not uncommon for engineers to associate pleasing or 'musical' lower-order even harmonics with valve-based equipment, while linking 'harsher' odd harmonic content to solid-state technology. It is true that older valve-based devices have a tendency to generate 'rich' even harmonic content, although it is important to note that

Harmonic Distortion (continued)

this is largely due to the topology of such equipment (class 'A').
'Near-clipping' or 'upper headroom' checks are a simple way to
investigate the suitability of harmonic and dynamic distortion to the
audio material and as a way to assess the impact of noise generated
at extremes of operation.

Microphone Preamplifier Noise Floor

The terminating of the input of a preamplifier, i.e. the connecting
or 'shorting' of pins 2 and 3 at the input stage, is a quick way to
determine the noise floor of a given device. Technicians should
ideally use a male XLR connector with a 200-ohm resistor wired in
series between the two pins (2 and 3) for this type of testing, as this
should present the preamp with the equivalent output impedance
of an 'average' dynamic microphone.

Variable Impedance

Some preamplifiers offer user modifiable control over input
impedance. Such control may be continuous or discrete (few
values) and it may have a significant impact over sound quality
in circumstances when high impedance microphones, e.g. some
ribbon models, are connected to the preamplifier via long cable
runs. In such cases, a preamp high impedance setting may be
recommended, as it may make signals seem 'bigger' or less 'thin'
(more low-end content). NB the altering of input impedance should
have very little impact on the signal of condenser microphones.

Suggested Microphone Preamplifier Evaluation Procedure

The process of evaluating the suitability of microphone
preamplification may appear arbitrary to recordists with less
experience, although it is possible for a simple methodical approach
to take the place of 'blind' trial and error.

Suggested Microphone Preamplifier Evaluation Procedure (cont.)

The following is a suggested vocal microphone preamp selection routine that may allow for a quick comparison between different devices:

1. Set up a given microphone preamplifier and DAW to record vocal signals through a well-known large diaphragm condenser microphone.
2. Set the input gain so that very quiet vocal sounds are amplified to reach or approach the standard operating level of the preamp.
3. Ask a vocalist or an assistant to say (or sing) percussive or transient rich words at very quiet levels, e.g. 'test', 'tick', 'tock', 'click', 'clock', 'pop', 'pup', etc.
4. Monitor the device's output signal ensuring that all transients and high-frequency content are preserved and that the noise floor is not too high.
5. Check that a sense of intimacy, i.e. the feeling that a person speaking/whispering into the listener's ear, is being captured.
6. Set the input gain of the preamp so very loud singing is amplified to a level approaching the device's clipping point.
7. Ask a vocalist or an assistant to sing sustained vowel-based material at very loud levels, e.g. ah, eh, oh, etc.
8. Check the added harmonic distortion for suitability.

This test can be particularly effective for the determining of whether a preamplifier is suitable for the recording of main vocals and it may also help establish the need for dynamic range processing, i.e. compression.

The first part of this procedure (steps 1 to 5) can be used for the comparison of preamps to be employed for the detailed recording of acoustic instruments. The second part (steps 6 to 8) should be effective for the comparison of preamps to be employed for the recording of loud, 'energetic' sources, e.g. electric guitar, snare drums, etc.

The following page contains a few examples of common microphone preamplifiers used in music production.

PREAMPLIFIER (EXAMPLES)

NEVE 1073-DPA

MILLENIA HV-3C

MANLEY DUAL MONO

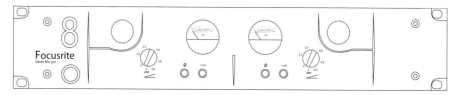

FOCUSRITE RED 8

EQUALISERS

Equalisation may be described as frequency (band) specific level control and it is commonly used to alter the timbre of audio signals or the response of sound reproduction systems. Equalisers are frequently utilised to repair or enhance recorded material, improving tone and clarity (balance), and are also used creatively to personalise music productions. The following is an overview of equaliser types and their applications.

Filters

Filters are primarily used for the removal of unwanted signal content, e.g. rumble, hiss, etc., providing an increasing level of attenuation above or below a given frequency. The rate of this attenuation is given in decibels per octave (filter slope) and it commonly varies between devices. Low and high pass filters may present no variable controls, i.e. incorporating fixed cut-off frequency and slope, or may offer selection facilities over:

- Frequency
- Slope (mostly found in digital units, whereas analogue filters commonly present fixed slopes).

Filter Slope

Filter slopes hold a direct proportionality with the number of reactive components (poles) used in EQ design:

- 1st Order / 1-Pole Filters: 6 dB per octave roll-off
- 2nd Order / 2-Pole Filters: 12 dB per octave roll-off
- 3rd Order / 3-Pole Filters: 18 dB per octave roll-off
- 4th Order / 4-Pole Filters: 24 dB per octave roll-off.

From the list above it is easy to infer that in filter design, the number of poles (or the filter order) multiplied by six equals the slope of attenuation in decibels per octave.

Filter Cut-Off Frequency

The 'cut-off' value specifies the frequency that will be attenuated by 3 dB when a filter is made active. This convention originates from basic crossovers, where such frequency represents the 'split' between the low and the high-ends of the spectrum (sent to the 'woofer' and the 'tweeter' simultaneously at an attenuated level).

CUT-OFF FREQUENCY

The low-pass filter below has its cut-off frequency set at 5 kHz.

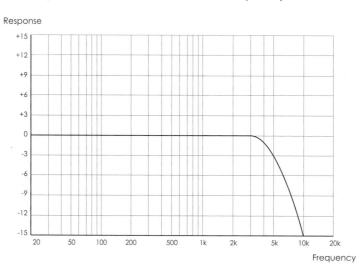

The high-pass filter below has its cut-off frequency set at 100 Hz.

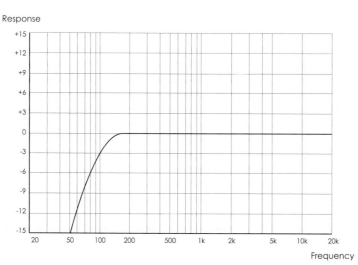

PASS FILTER CONTROLS

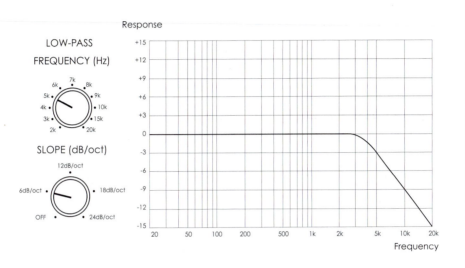

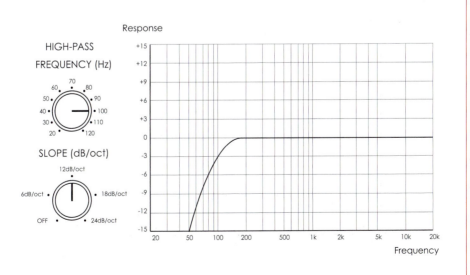

Pass-Filters and 'Rumble'

High-pass filters are generally used to remove 'rumble' from signals. These undesirable low-frequency sounds may originate from air-conditioning units, traffic, vibration from microphone stands, etc.

Mixing console high-pass filters commonly present cut-off frequencies between 85 and 150 Hz and when employed incorrectly may lead to the removal of important musical content from instruments such as bass guitar, bass drum, bass tuba, piano, etc. The use of pass filters should in such cases be avoided or approached with caution.

Shelving Equalisers

Shelving equalisers allow for (practical) constant attenuation or boost above (high-shelving) or below (low-shelving) a given frequency. Shelving EQ labelling methods vary considerably between manufacturers, although it seems sensible to adopt the 'stop frequency' as the nominal frequency of a given device (as AMS Neve and other manufacturers do), i.e. the frequency where boost or attenuation has reached a maximum 'stable' level.

Budget shelving units offer simple control over gain, commonly labelled as 'bass' or 'treble', while professional devices offer selection over frequency and gain.

Shelving EQ and Pass Filters

Shelving equalisers should be used with great caution and ideally employed alongside pass filters. As an example, it is vital to remove rumble before using a low-shelving equaliser to boost low frequencies.

Shelving EQ Emphasis and De-Emphasis

Shelving equalisers were (and still are) widely used in analogue tape-based recording sessions. Due to the noise floor of the medium, it is common for engineers to boost or emphasize the high-end of the material during the recording stage, reversing the process (attenuating the high-end) during mixdown. The 'de-emphasis' applied during mixing returns the audio content to its original timbre while reducing the audibility of tape hiss.

LOW-SHELVING CONTROLS

Response

GAIN (dB)

FREQUENCY (Hz)

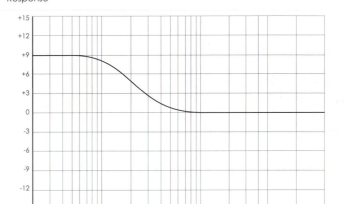

Frequency

Response

GAIN (dB)

FREQUENCY (Hz)

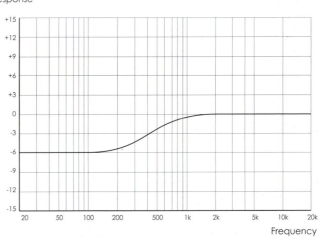

Frequency

HIGH-SHELVING CONTROLS

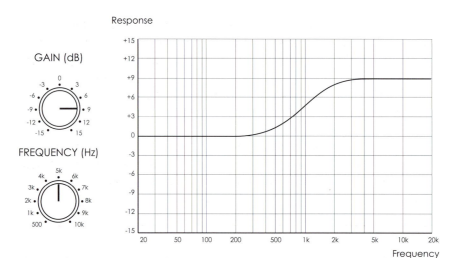

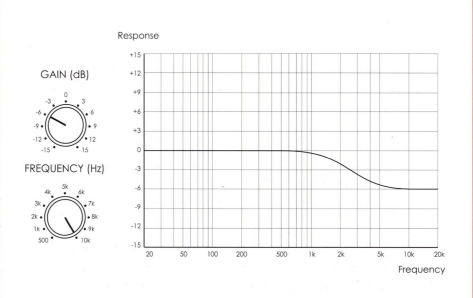

Peaking / Peak Equalisers

Peak equalisers, sometimes also referred to as 'bell' equalisers, are usually found affecting the mid-range band of 'budget' or semi-professional consoles. This type of EQ offers variable control over gain, but no frequency or bandwidth selection facilities.

Bandwidth

The value known as bandwidth describes the main area of effect (or the 'focus') of an equaliser. This value may be given in octaves or it may be presented as a single number known as Q or 'quality factor' (described in more detail later in this book). The bandwidth of a 'peak' equaliser lies between the two points found at three decibels above maximum attenuation or below maximum boost.

Digital devices, including plug-ins may offer extremely narrow Q factors or bandwidths, which may not be reached by analogue hardware devices. This allows for the very precise removal of small bands of offending frequencies from signals.

Symmetrical and Non-Symmetrical Equalisers

Symmetrical equalisers present identically shaped curves for cuts and boosts of equal absolute values, e.g. + 6 and – 6 dB. Non-symmetrical equalisers tend to offer wider boosts and narrower cuts, as the latter are frequently used for focused or 'surgical' repairs while boosts are approached more broadly (see pages 62 and 63).

Constant and Non-Constant Bandwidth Equalisers

Some EQ units present a constant quality factor that is independent of gain, while others offer a non-constant 'Q', which changes according to the depth of boost or cut, i.e. the greater the equaliser gain applied the narrower the range of the effect. As examples, SSL 'E' series equalisers present constant Q and are favoured for precise boosts or cuts, while the same company's 'G' series EQ is considered to be less surgical or more 'musical' by some, as it offers a non-constant, gain-dependant bandwidth (see page 64).

'SYMMETRICAL' PEAK EQ CONTROL

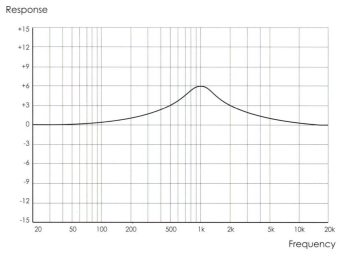

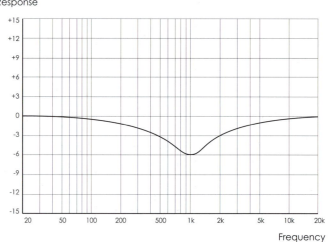

'ASYMMETRICAL' PEAK EQ CONTROL

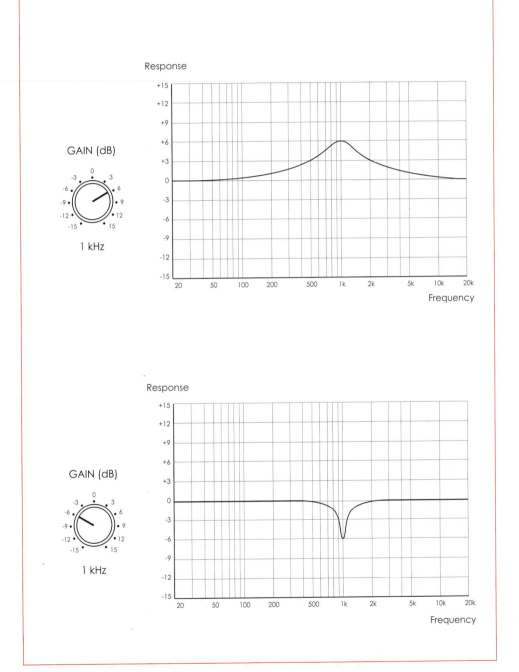

CONSTANT / NON-CONSTANT 'Q' PEAK EQ

Constant bandwidth EQ

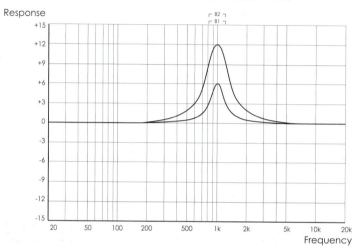

Non-constant bandwidth EQ
(gentler boost is wider in effect)

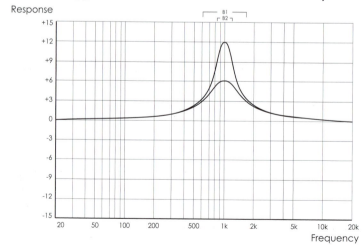

Graphic Equalisers

Graphic equalisers are effectively a series of 'peak' EQ units that control overlapping bands, arranged under a user-friendly physical layout. High-end graphic equalisers commonly offer a minimum of 31 bands, spaced at a third of an octave intervals.

Users should avoid the boosting and cutting of adjacent bands of graphic EQ, as this may defeat the purpose of the processing and introduce phase-related problems.

'Sweep' Equalisers

Also commonly referred to as 'bell' equalisers, sweep EQ units are similar to 'peak' devices, while offering added control over frequency selection. NB 'sweep' equalisers do not offer control over bandwidth.

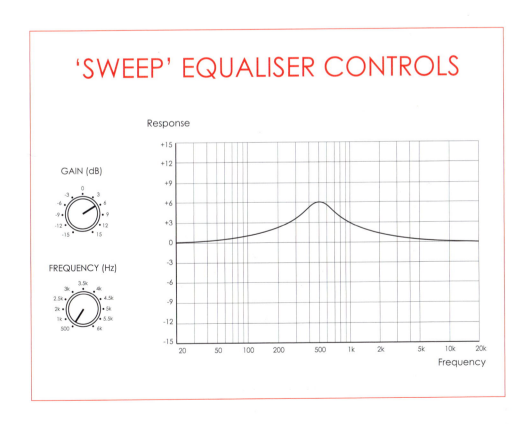

'SWEEP' EQUALISER CONTROLS

MUSIC PRODUCTION: RECORDING

Semi and Fully-Parametric Equalisers

Parametric equalisers allow for user selection of frequency, gain and bandwidth or Q. If the control over Q is continuous, devices are said to be of fully parametric design, while if the same control is non-continuous or stepped, they are labelled as semi-parametric.

The Q Factor

The Q (quality) factor is used to describe the area of action of an equaliser. In a somewhat counter-intuitive fashion, the higher the value of Q the narrower the effect of a device, i.e. the more precise or 'surgical' the equalisation.

Narrow Qs are commonly used to remove small, specific portions of a signal, e.g. the unsympathetic resonance of a snare drum, without affecting the rest of the sound of the instrument.

The following are approximate values for Q that are worth remembering:

Q = 0.67 (two octaves)
Q = 1.41 (one octave)
Q = 2.87 (1/2 octave)
Q = 4.31 (1/3 octave)
Q = 17 (one semitone)

It may be helpful to consider the Q value of 1.41 (a bandwidth of one octave), as the point that separates 'narrow' and 'wide' EQ effects, i.e. equalisation will be focused if $Q \geq 1.41$, while it will be broader when $Q \leq 1.41$.

Paragraphic Equalisers

Paragraphic equalisers may be described as graphic equalisers that offer added control over bandwidth (alongside gain).

SEMI-PARAMETRIC CONTROLS

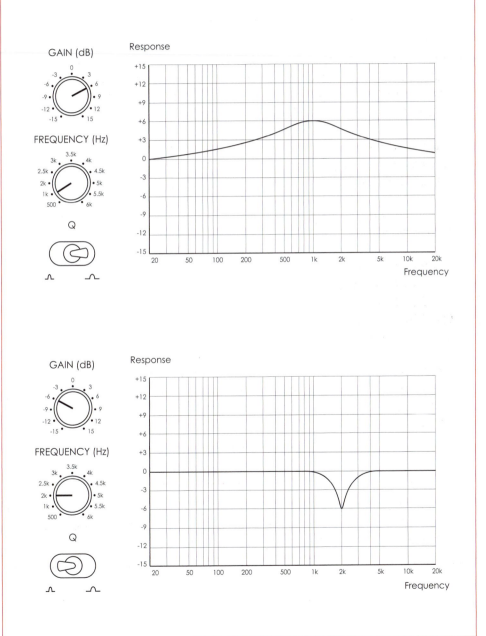

FULLY-PARAMETRIC CONTROLS

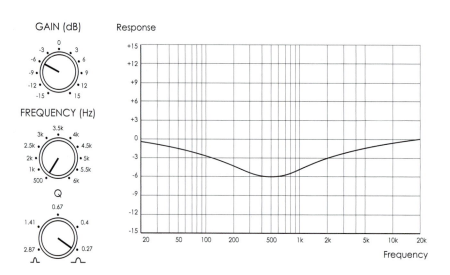

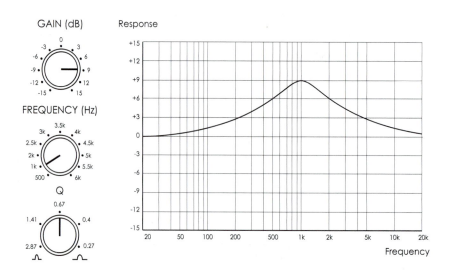

EQUALISER (EXAMPLES)

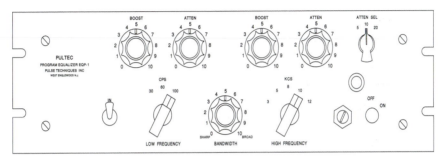

PULTEC EQP-1

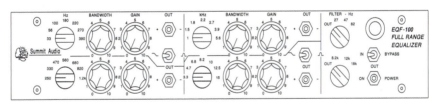

SUMMIT AUDIO EQF-100

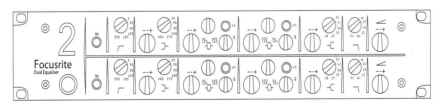

FOCUSRITE RED 2

EQUALISER (EXAMPLES)

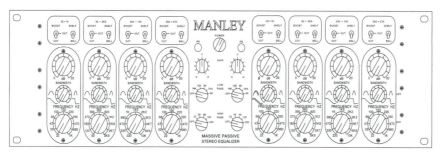

MANLEY MASSIVE PASSIVE

CHANDLER GERMANIUM TONE CONTROL

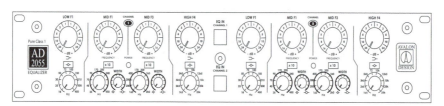

AVALON AD2055

EQUALISER (EXAMPLES)

API 550

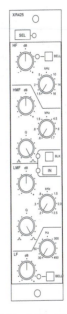

SSL XR425

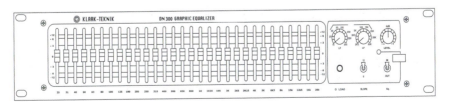

KLARK-TEKNIK DN300

EQUALISER (EXAMPLES)

PRO TOOLS' EQ 3

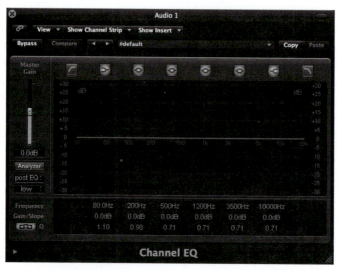

LOGIC'S CHANNEL EQ

DYNAMIC RANGE PROCESSORS

Dynamic range processors may be broadly described as gain riding devices, providing automatic control over the level of signals. This control may be effectively used to reduce (compressors and limiters) or extend (expanders and gates) the dynamic range of audio material.

Compressors / Limiters

Compressors / limiters are used to reduce the dynamic range of signals and may incorporate all or some of the following user-adjustable parameters:

- Threshold – The level at which a compressor will initiate gain reduction, i.e. the level at which signals will start being affected by a compressor (possibly influenced by a 'knee' function).
- Ratio – A gain reduction 'intensity' control (x:1), this parameter determines the proportion between input and output levels above the threshold (overshoot).
- Attack – A time-related function determining the speed at which a compressor will reach its user-defined ratio of gain reduction (x:1) once input signals rise in level above the threshold. The primary function of the attack parameter is to determine the speed at which a compressor will start working once signals overshoot the threshold (although attack still applies for further increases in level above the threshold).
- Release – A second time-related function determining the speed at which a compressor will diminish its gain reduction when input signals fall in level from previous readings above the threshold. The primary function of the release parameter is to determine the speed at which a compressor will stop affecting signals once their level drops below the threshold (although release will also affect decaying signals above the threshold).
- Knee – This parameter may be described as a fine-tune control over threshold and attack, i.e. a soft knee function effectively lowers the threshold and slows down the attack of a compressor.
- Make-up gain – A post-signal processing level control commonly found at the output stage of compressors.
- Input level – On devices with no time constant controls, i.e. no attack or release, this function dictates the extent of compression by raising or lowering the level of input signals above or below a fixed threshold.
- Output level – On devices with no time constant controls, this function dictates the overall output level.

Although the gain-reduction cell corresponds to a portion of a larger, complex system, compressors are usually described according to their gain stage, falling (mostly) within the following categories (see page 78):

- Delta-Mu (Vari-Mu)
- Opto-Electrical
- FET
- VCA

Compressors and Microphone Technique

Skilled singers use microphone techniques to help avoid the need of compression during recording. Such techniques may be as simple as changing distances to the microphone dynamically or accompanying short bursts of high-impact vocal delivery with body movements that make the performer 'look away' from the microphone, e.g. the raising of arms, a slight spinal bend backwards, etc.

Fader Riding and Compression

Manual compression was used extensively in the heyday of analogue tape recording, where due to the limited dynamic range of the medium, engineers were commonly required to 'ride faders' to tape. As an example, it was usual for classical music recordists to follow the score slightly ahead of the performers in order to prepare themselves for fortissimo sections that would require faders to be brought down in level (in order to avoid 'clipping'). The aforementioned process was generally followed by a mixing stage were manual expansion would return the audio program to its original dynamic range.

Compressors may be thought of as an automated replacement to be used when human reaction times would be too slow for appropriate fader riding.

Page 75 depicts a comparison between the effects of compression and those of fader riding.

COMPRESSORS AND GAIN

Original Signal (Compressor Bypassed)

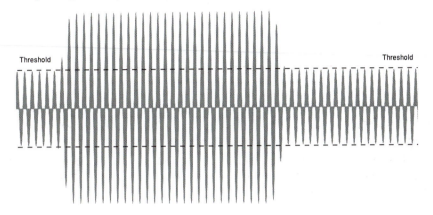

Threshold

Threshold

Compressed Signal

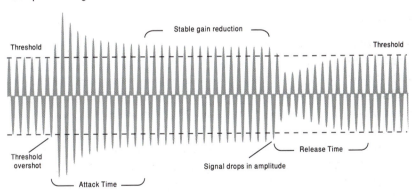

Stable gain reduction

Threshold

Threshold

Threshold overshot

Attack Time

Signal drops in amplitude

Release Time

Equivalent manual fader riding

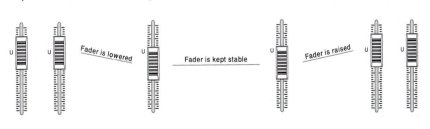

Fader is lowered

Fader is kept stable

Fader is raised

Peak / RMS Detection

Compressors may react according to the peak or the RMS of input signals. Level detection circuits based on peaks present a faster response and are usually more efficient in the control of extreme transients (limiting). RMS-based dynamic processors work on a level average and tend to be gentler in effect, although they may be unsuitable for percussive material.

'Look-Ahead' Detection

'Look-ahead' circuitry is commonly used when instantaneous gain reduction is needed, e.g. for device protection, broadcast limiting, etc. Compresssors and limiters that offer a 'look-ahead' function, delay the signal in their main audio path so level detection may take place and gain reduction initiated before signals overshoot the threshold.

Limiters

Limiters may be broadly described as compressors with very high ratios and fast attack (and ideally with a look-ahead function). These are commonly used to maximise the loudness of audio material, to protect equipment and for broadcasting purposes.

The Side-Chain

Compressors incorporate two audio paths, which are commonly fed with the same input signal. The function of the first one, the side-chain, is that of level detection, i.e. this is where the amplitude of the program material is determined and the audio signal (AC) is converted onto a control voltage (DC). The second and main path is where signal processing (compression) takes place.

A compressor's side-chain may be fed with a signal that differs from the one that will be processed in conditions where the desired

The Side-Chain (continued)

gain reduction may not be achieved easily. This could be the case when the amplitude of the portion of the signal to be compressed is not sufficiently greater than the material's average or when gain reduction is triggered by components that should not be attenuated, e.g. the low-end.

De-Essing

De-essing is a common alternative when sibilance (high amplitude consonants such as 's', 't', etc.) cannot be controlled through the use of pop shields. A de-esser is essentially a compressor with an equalised side-chain input.

In de-essing, a main vocal signal is fed to a compressor's input and an exaggeratedly sibilant (equalised) version of the same vocal is fed to the device's side-chain. The latter ensures gain reduction only takes place, or it is maximum, at points when sibilance is detected.

A process similar to de-essing may be used to reduce fretboard and finger 'squeaking' sounds from acoustic and electric guitar recordings.

DE-ESSER

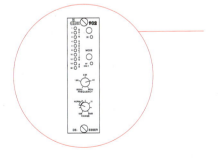

dBx 902

COMPRESSOR TOPOLOGY

DELTA-MU

Limited distortion-free gain reduction range
Program-dependant attack / release times
Commonly used for buss compression
Ratio dictated by program levels
Non-linear attack / release times
Limited controls (no ratio)
Slowest attack times

Examples: Fairchild 670, Manley Vari-Mu

OPTO-ELECTRICAL

Program-dependant, non-linear gain reduction
Limited distortion-free gain reduction range
Fixed threshold (Input and Output controls)
Decelerating (musical) release times
Slow attack times
Soft knee

Examples: LA-2A, LA-3A

FET

Extended user-definable controls, e.g. ratio
Harmonic artifacts considered musical
Less restricted gain reduction range
Favoured for extreme compression
Fast attack and release times
Potentilly high noise floor

Examples: Urei 1176LN, Daking FET II

VCA

Greatest range of gain reduction
Fastest attack and release times
Most linear gain reduction
Varied in quality

Examples: dbx160, SSL G-Series

COMPRESSORS (EXAMPLES)

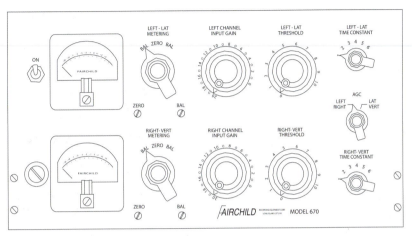

FAIRCHILD 670

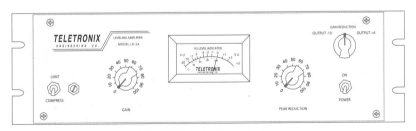

TELETRONIX LA-2A

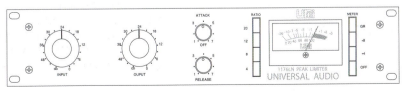

UREI 1176-LN

COMPRESSORS (EXAMPLES)

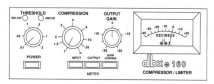

DBX 160

API 525

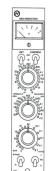

NEVE 2264

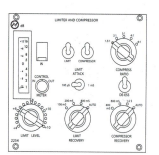

NEVE 2254

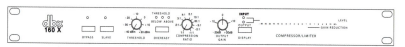

DBX 160X

SSL G384

EMPIRICAL LABS DISTRESSOR

COMPRESSORS (EXAMPLES)

PROTOOLS' COMPRESSOR / LIMITER DYN3

LOGIC'S COMPRESSOR

Expanders / Gates

Expanders / gates are used to increase the dynamic range of signals and may incorporate all or some of the following user-adjustable parameters:

- Threshold – The level at which an expander / gate will stop gain reduction, i.e. the level at which a gate will 'open'.
- Ratio – A gain reduction 'intensity' control (x:1), this multiplier determines the proportion between input signal undershoot (the portion of the signal below the threshold or below threshold minus hysteresis) and effective output level.
- Attack – A time-related function determining the speed at which an expander / gate will stop gain reduction completely (gate fully open).
- Hold – The length of time in which an expander / gate will remain inactive, i.e. delay its gain reduction, once signals undershoot the threshold (or threshold minus hysteresis).
- Release – A second time-related function determining the speed at which an expander / gate will realise its full ratio or range of gain reduction (gate fully closed).
- Range – The maximum level of gain reduction applied to gated signals.
- Hysteresis – A decibel value describing the difference between the threshold (the point at which the gate will open) and the point at which a gate will start attenuating signals (gate closing).

Gating and Expansion

Expansion implies variable gain reduction that is proportional to a ratio, while gating entails a fixed, predetermined level of attenuation, i.e. a given amount of decibels, taken from signals that do not overshoot a set threshold.

The following is an illustration comparing the effects of gating with those of fader riding.

GATES AND GAIN

Original Signal (Gate Bypassed)

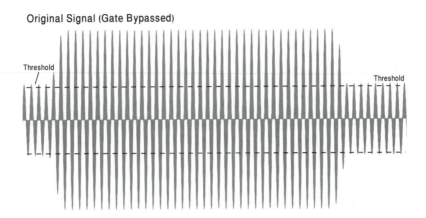

Threshold

Threshold

Gated Signal

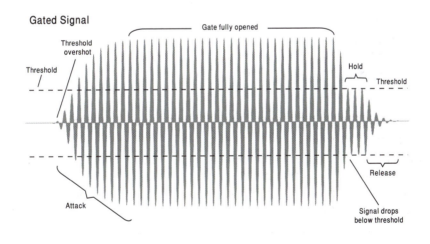

Gate fully opened

Threshold
overshot

Threshold

Hold

Threshold

Release

Attack

Signal drops
below threshold

Equivalent manual fader riding

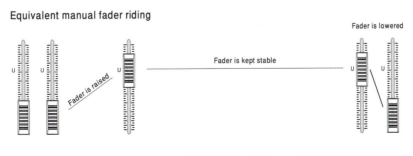

Fader is lowered

Fader is kept stable

Fader is raised

Expanding Signals

Expansion is commonly used to reduce the level of decaying signals, e.g. expansion may help to reduce the audibility of amplifier noise, as an electric guitar's signal falls back to very low levels (approaching the level of idle amplifier 'hum').

Duckers

Duckers are commonly used in broadcasting when a main signal, e.g. a voice-over, controls (expands / attenuates) the level of another unrelated one, e.g. background music.

Key input

In a similar fashion to side-chained compression, key input gating allows for the dynamic range of a signal to be controlled by the level of another, possibly unrelated one. As an example, a drum 'room' microphone may be gated with a snare drum signal as a key input, so that only when the snare is struck, the sound of the room microphone will be audible.

GATES (EXAMPLE)

DRAWMER DS201

GATES (EXAMPLES)

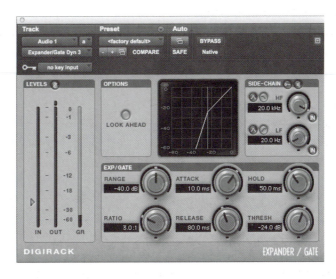

PROTOOLS' EXPANDER / GATE DYN3

LOGIC'S NOISE GATE

MIXING CONSOLES

Audio consoles vary greatly in design and features while sharing the same basic functions, commonly working as:

- Central routing devices – Allowing for signals to be routed between different (own) internal and external points with minimal need for patching, e.g. from microphone input to multitrack send.
- Signal processing and amplification devices – Offering facilities for level matching and signal manipulation, e.g. equalisation.
- Signal summing devices – Making it possible for multiple signals to be summed or combined.

Console Types

Mixing consoles may be classified according to design, falling into one of the following categories:

- Line mixer – A very basic device with a single set of I/O (input / output) module faders and no microphone preamps. Line mixers are commonly used for the summing of line level signals, e.g. the back end of DAW-based mixing systems.
- Split console – A mixer that accepts line and microphone level signals, while still offering a single set of I/O module faders. These are commonly operated with half the number of channels sending signals to the multitrack recorder, while the other half is used for monitoring (multitrack returns).
- Semi in-line console – A console that accepts line and microphone level signals and offers two sets of faders per module, corresponding to the two different signal paths (channel and monitor or Mix B). Semi-in line consoles commonly present a 'tape input' that may be routed to monitor or Mix B paths (record) or to channel paths (mixdown).
- In-line console – Similar to a semi in-line console, although a true in-line mixer has the output of its corresponding multitrack machine feeding both (channel and monitor) paths simultaneously. The use of a patchbay is usually required for the operation of this type of console.
- Matrix console – Commonly used for live sound stage monitor mixing, matrix consoles may offer a considerable number of auxiliary busses, but no linear channel faders (not commonly used for recording purposes).

The illustrations that follow depict basic set-ups incorporating the different types of mixing consoles described in this section.

LINE MIXER

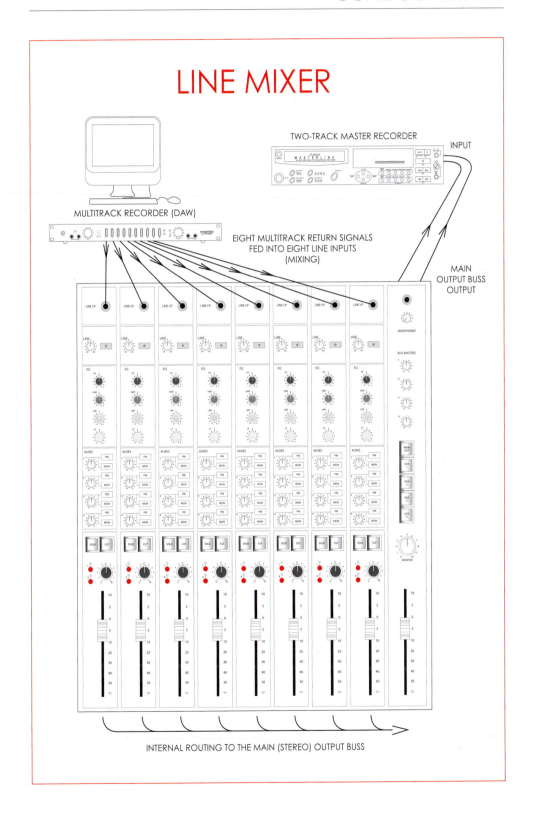

TWO-TRACK MASTER RECORDER

INPUT

MULTITRACK RECORDER (DAW)

EIGHT MULTITRACK RETURN SIGNALS
FED INTO EIGHT LINE INPUTS
(MIXING)

MAIN
OUTPUT BUSS
OUTPUT

INTERNAL ROUTING TO THE MAIN (STEREO) OUTPUT BUSS

SPLIT CONSOLE

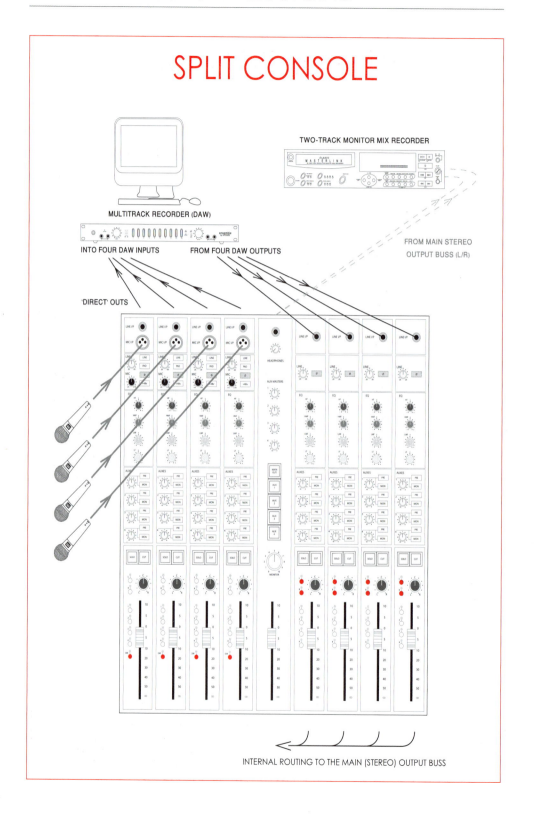

TWO-TRACK MONITOR MIX RECORDER

MULTITRACK RECORDER (DAW)

INTO FOUR DAW INPUTS   FROM FOUR DAW OUTPUTS

FROM MAIN STEREO

OUTPUT BUSS (L/R)

'DIRECT' OUTS

INTERNAL ROUTING TO THE MAIN (STEREO) OUTPUT BUSS

IN-LINE CONSOLE

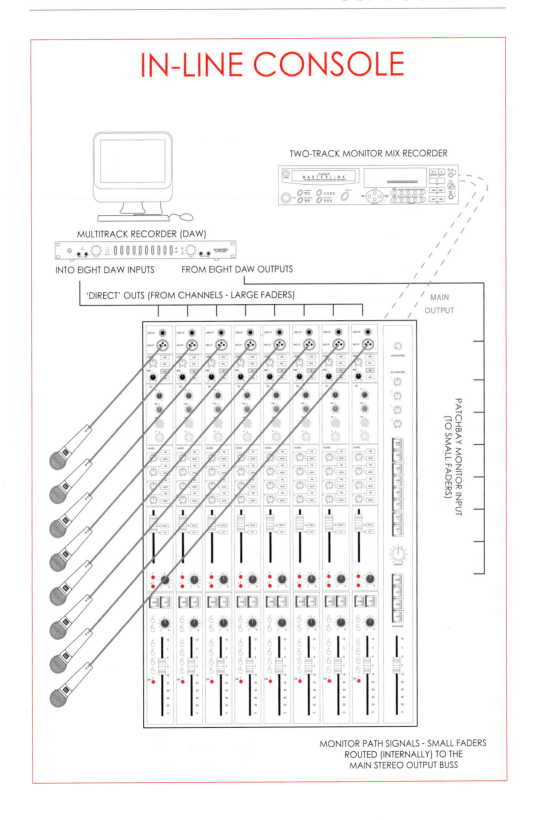

TWO-TRACK MONITOR MIX RECORDER

MULTITRACK RECORDER (DAW)

INTO EIGHT DAW INPUTS   FROM EIGHT DAW OUTPUTS

'DIRECT' OUTS (FROM CHANNELS - LARGE FADERS)

MAIN
OUTPUT

PATCHBAY MONITOR INPUT
(TO SMALL FADERS)

MONITOR PATH SIGNALS - SMALL FADERS
ROUTED (INTERNALLY) TO THE
MAIN STEREO OUTPUT BUSS

Console Elements

Mixing consoles may be very broadly depicted as a collection of channels and busses interconnected via routing with flexible monitoring capabilities. Channels commonly carry signals originating from individual sound sources, e.g. a microphone, while busses are summing points where multiple signals are combined. This allows for numerous sound sources to feed a single input device, e.g. two bass drum microphones routed to the same track, guitar and snare drum signals sent to the same reverb unit, multiple signals summed and sent to a headphone amplifier, etc. It is important to note that the monitor or Mix B paths of in-line consoles may be simply described and treated as alternative channels that carry line level signals that commonly originate from recording devices. This may be easier to understand when one works with a digital console, where signals 'going to' and 'returning from' the recorder are usually assigned to channels on different layers, e.g. Channels 1 – 24 for 'send' and Channels 25 – 48 for 'return'. In the case of analogue consoles, these two 'layers' translate as two faders (or a fader and a rotary potentiometer) on the same I/O module.

The following sections contain a brief description of common mixing console elements and their functions. It is important to note that some digital mixers may offer extra facilities such as analogue-to-digital and / or digital to analogue conversion, internal effects processing, etc.

Input / Output (I/O) Modules or Channel Strips

Common I/O modules or channel strips incorporate the following sections:

1. Input Section

The input section of a module corresponds to the entry point to its channel path. This section may include the following controls:

- 'Mic' gain – Microphone preamplifier level control
- Line trim – Line amplifier level control
- Input flip – Channel input selection (microphone or line)
- Sub-Group – This switch assigns the signal from a group buss to its corresponding channel path, e.g. group buss 1 will feed channel 1, etc.
- 48 V – Phantom power
- Pad – Fixed microphone input level attenuation (– 20 to – 30 dB)
- Polarity (also referred to as 'phase') – Used to invert the polarity of signals
- High-pass filter – Commonly with fixed cut-off frequency if placed here.

Shared Microphone / Line Input Preamplifiers
Budget consoles commonly offer a single gain control for both microphone and line inputs. In such cases, line level signals are attenuated (internally) through a resistor network, before reaching the preamplifier.

---

2. Auxiliary Section
The auxiliary section of the I/O module provides access to the auxiliary busses commonly used for effects sends, headphone mixes, etc. This section may contain some or all of the following controls:

- Source selection, e.g. large or small fader / channel or monitor (Mix B) path
- Pre or post-fader sourcing selection
- Send to auxiliary on/off
- Auxiliary send level controls (each corresponding to a given mono or stereo buss)
- Auxiliaries pairing switch (which may turn a level control into 'pan').

Pre / Post Auxiliary Sourcing
'Cue' mix signals are normally sourced pre-fader and pre-cut, so engineers may change the balance and content of their control room (monitor) mix without affecting the musicians' headphone levels. This includes the 'sends' and 'returns' to and from processing units, if these are used to add effects to the 'cue' mix, e.g. reverb added to a singer's headphone signal.

In mixdown, 'auxiliaries' used for effects sends are commonly sourced post-fader and post-cut, where the 'muting' or 'cutting' of a 'dry' signal (instrument) will be followed by a corresponding 'mute' or 'cut' of its effects returns or 'wet' version.

Auxiliary Busses Feed to Group Busses / 'EFX'
Some consoles allow for the routing of auxiliary buss outputs to the group busses. This facility may be explored for the creation of multi-

---

Auxiliary Busses Feed to Group Busses / 'EFX' (continued)
channel or 'surround' mixes, as long as each group buss is fed to a corresponding amp/speaker. In such case, the I/O module 'aux' level pots control the amount of channel 'send' to the different console outputs, e.g. channel to Aux 1 to Group 1 to left front speaker. Consoles with stereo auxiliary busses may also offer a panning function facilitating the routing of signals between speakers.

The same facility may alternatively be employed for the real-time use of computer effects through traditional 'aux send' means, in set-ups where the DAW is fed via group busses.

3. Insert Section

Inserts are commonly used as a means to access outboard equipment for the serial processing of signals that can be improved or enhanced, e.g. inserts may be used to equalise vocals that lack 'sparkle' or 'air', where the original 'dull' microphone sound is enhanced before it is recorded.

Insert sections should ideally offer:
- Channel or monitor path (Mix B) assignment selection
- Pre or post-EQ placement option
- An 'insert in' or 'bypass' function, allowing operators to compare the original and the processed versions of a signal.

4. EQ Section

Professional mixing consoles frequently incorporate multi-band equalisers and cut filters in their I/O modules. The type of equalisation offered may vary according to frequency band, where shelving or sweep components are commonly used at the two ends of the spectrum (low and high bands) and semi or fully-parametric ones manipulate the mid-range bands (low and high-mid). EQ sections should ideally offer channel or monitor path assignment selection and a 'bypass' function, allowing operators to quickly A/B the original and the equalised versions of signals.

5. Dynamics Processing Section

High-end mixing desks may feature I/O module and output buss compressors, limiters, expanders, gates and duckers. In the case of in-line consoles, access to the dynamics section may be assigned to channel or to monitor (Mix B) paths.

The dynamics processors found in I/O modules may offer:
- Assignment selection (channel or monitor path)
- Placement options (pre or post-EQ)
- Side-chain or key input access
- A 'dynamics on/off' or a 'bypass' switch.

6. Fader Section(s)

The large fader section of a module is allocated to the channel path, while the small one is assigned to the monitor or Mix B path. This always holds true, regardless of global operation status (Record or Mix), unless a console is capable of operating in 'fader reverse'. Fader sections commonly contain:
- A large or a small fader
- A solo switch
- A mute switch.

Destructive / Non-Destructive Solos

Professional consoles offer two types of solo facilities, namely destructive (or in-place) and non-destructive (split into AFL and PFL).

1. Destructive (In-Place) Solos

When a destructive solo is performed on a path, all other similar (or all other) paths are cut / muted, i.e. only the 'soloed' path will have its routing preserved. Destructive solos are commonly used in mixdown sessions when engineers aim to showcase a single channel or element, which will feed the main output stereo buss in isolation, e.g. a lead vocal acapella section, a rhythm guitar break, etc.

2. Non-Destructive Solos

When a non-destructive solo is performed, a copy of the signal in the soloed path is routed to a dedicated solo buss, which is automatically sourced at the control room monitoring section. All other paths and their routings are not affected. Non-destructive solos are commonly used in recording sessions, allowing for the monitoring of signals at different stages of the chain. This feature can be particularly helpful for troubleshooting.

There are two common types of non-destructive solos:
AFL / After Fader Listen – The solo buss feed is post-fader.
PFL / Pre Fader Listen – The solo buss feed is pre-fader.

Solo in Front

Some consoles offer a solo in front feature. A solo in front is essentially a variation of an AFL non-destructive solo, allowing for the monitoring of all 'non-soloed' paths as a 'background' (at lower level in relation to the main soloed path). The amount of dimming applied to the 'non-soloed' paths is given in percentage form, e.g. 50 percent implies that the non-soloed paths will be monitored at half the level of the soloed path.

Solo Safe / Solo Isolate

A path may be set onto 'solo isolate' or 'solo safe' mode. This implies that such path will not perform solos destructively and, more importantly, that such path will not be cut in the case of a destructive solo, e.g. this feature is commonly used to ensure that channels fed by effects processor returns will not be muted if their counterpart effects processor send paths are soloed during mixdown.

7. The Routing Matrix

Routing matrices offer access to group and / or main output busses. Some professional consoles contain two matrices in each module, while others may only incorporate a single one, commonly placed within or near the large fader section.

The use of routing matrices varies according to console status. In 'Record', status group busses are commonly used to feed the multitrack recorder, while in 'Mix', group busses are frequently used as a means to access outboard processing equipment (working as 'extra aux busses'). Routing matrix sections usually incorporate a pan pot for odd/even assignment.

Pan Assign

Pan pots are used in conjunction with routing switches to address signals to multiple destinations. As an example, engineers choosing to record multi-microphone drum set-ups to two tracks of a multi-

Pan Assign (continued)

track recorder (stereo) may assign the output of the different channels to two (odd/even) group busses, choosing to pan left or right in order to establish a stereo image during playback.

Example – Recording Drums in Stereo (Drummer's Perspective):

- Bass drum – Routed to groups 1 and 2 (panned centre)
- Snare drum – Routed to groups 1 and 2 (panned to 11:00)
- Hi-hat – Routed to groups 1 and 2 (panned to 10:00)
- OH left (hi-hat) – Routed to groups 1 and 2 (panned hard-left)
- OH right (floor tom) – Routed to groups 1 and 2 (panned hard-right).

Pan vs. Balance

Pan and balance pots precede routing decisions. The first allows for a single, mono source to be routed to pairs of destinations at different levels, e.g. when a mono signal is routed to the left and the right side of a Mix or Output Buss, etc., while the second allows for the control over the level of the two 'sides' of stereo signals, when routed to stereo destinations.

PANNING

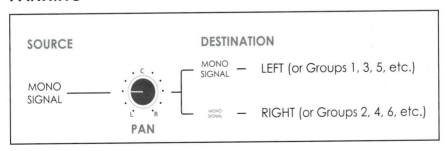

Pan vs. Balance (continued)

BALANCE

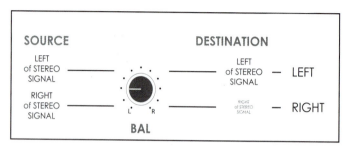

- It is common for consoles to incorporate pan pots in mono channels and balance pots in stereo channels.
- Balance control pots are commonly found in hi-fi amplifiers.

Direct Outputs

Some consoles offer direct output facilities, which allow individual signals to be routed to 'multitrack sends' directly, i.e. without going through group busses. Sophisticated professional consoles allow operators to choose where the 'direct' signal is sourced from, e.g. channel path pre-fader and pre-processing (EQ and dynamics), channel path pre-fader, but post-dynamics or channel path post-fader.

A few mixing desks allow for the routing of monitor path signals to 'multitrack send'. This may be used for bouncing tracks or for maximising the recording capabilities of smaller consoles. It is very important to note that feedback loops may occur easily between the monitor paths and the recorder, e.g. if the 'direct' function is sourcing the 'multitrack send' (off-buss) signal from the monitor path.

The following is an illustration depicting 'direct' and group buss multitrack assigning:

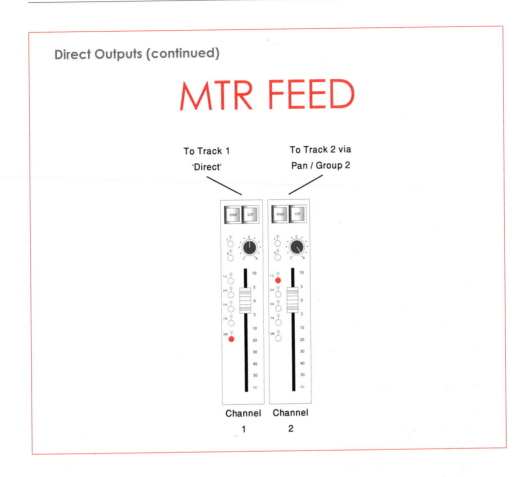

Direct Outputs (continued)

MTR FEED

To Track 1 'Direct'

To Track 2 via Pan / Group 2

Channel 1   Channel 2

The Master Section

The master section of audio consoles commonly incorporates all or some of the following elements:

1. Master Group Section

This section may offer linear faders controlling the level of group busses. A small routing matrix and a pan pot may also be incorporated, allowing for the output of group busses to be routed to the main stereo output buss. Some mixers also offer access to the auxiliaries from the group busses and this facility may be explored for the creation of streamlined 'cue' mixes based on instrumental groups, e.g. drum groups, guitar group, etc.

2. Master Stereo (Mix / Output) Buss Section

A linear fader is frequently used to control the main stereo output buss level. In some cases, two separate faders are found controlling the individual level of the left and right channels of the stereo buss. The master output buss section may also contain a 'mute' button.

3. Auxiliary Master Section

This section contains the master controls over the auxiliary busses, commonly including:

- Auxiliary buss master on/off switch
- Auxiliary buss master level
- Auxiliary buss master EQ (commonly shelving or cut-filters)
- Auxiliary pairing switch.

4. Effects (Echo / Rev) Returns Master Section

This section contains the master controls over the 'effects' or auxiliary returns and it may include:

- Effects returns master level control
- Effects returns master on/off switch
- Effects returns master EQ (commonly shelving or pass-filters)
- Routing matrix – Allowing for the assigning of effects to the main output buss and/or the headphone mixes.

5. Tone Generator / Oscillator Section

This section provides tone generating facilities, allowing for equipment testing, alignment and troubleshooting.

6. Meters

The master or central area of consoles may incorporate output, group and auxiliary buss meters alongside control room (two-track) metering facilities. The meter section may also include a global I/O module metering selection matrix.

7. Control Room Monitoring Section

The control room monitoring section may include:

- Monitoring level control – Main level to amplifier or active speakers
- Mono function – Left and right output buss combining function (used for checking polarity coherence)
- Dim – A monitor level attenuation switch
- Monitoring source matrix – Selection of control room monitoring sources, including the Mix Buss, auxiliary busses, 'cue' mixes and external inputs
- Speaker selection.

8. Communications Section

This section offers control over the communication between the control room and the studio (live room). Commonly referred to as the 'Talkback' section, this area may include:

- A built in microphone
- Level control
- Talkback and return talkback on/off switch
- Talkback and return talkback level controls.

Console Signal Flow (Block) Diagrams

The path that connects a transducer to the recording device may vary in size and, consequently, in number of gain stages. It is extremely important for technicians to have an understanding of signal flow, as without such knowledge it would be difficult, if not impossible, to optimise signals before committing them to the storage medium. Audio chains can be as basic as a single microphone connected to a computer interface with a built-in preamplifier or as complex as what is found in commercial multitrack studios built around large format consoles. In all cases, recordists must consider the variables that may affect sound quality and ensure that the path to the recorder is as short as possible and free of unreasonable gain changes in order to maximise the signal-to-noise ratio.

Home studio recordists who work with single sound sources in isolation commonly have little to worry about regarding signal flow, i.e. their signal paths are short, with one gain stage corresponding to the audio interface's 'mic' preamplifier. The picture is considerably different when engineers choose to record entire groups simultaneously using a mixing console and outboard processing. In such cases, the use of block diagrams may be essential as such documentation frequently provides technicians with a clear and linear overview of signal flow from console input to multitrack recorder to two-track recorder to loudspeakers, etc.

Signal flow diagrams come in different shapes and sizes, utilising a variety of symbols that may change according to manufacturer. However, a few universal rules seem to apply and are observed in the majority of documents, such as:

- Signals flow from the left to the right of a page.
- Routing destinations are depicted as vertical lines.

The following pages present an overview of block diagrams:

ROUTING A SIGNAL TO THE MULTITRACK RECORDER

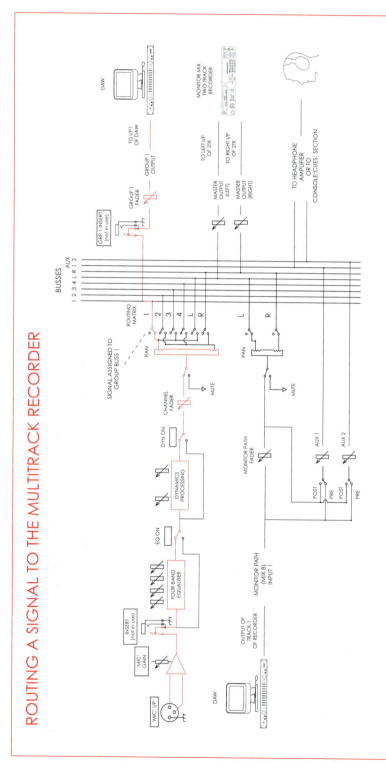

In this example a microphone signal is fed onto the channel path of a console being used as the front end for a DAW. The signal bypasses the insert section and it is processed by a four-band equaliser and a compressor (dynamics). After travelling through the channel fader and a pan pot the signal reaches a routing matrix where it is assigned to group buss 1. The output of this group is normalled to multitrack send 1, i.e. it is fed onto line input 1 of the audio interface, which should correspond to a track of the DAW.

ROUTING THE DAW RETURN SIGNAL TO THE MONITOR MIX (TWO-TRACK) RECORDER

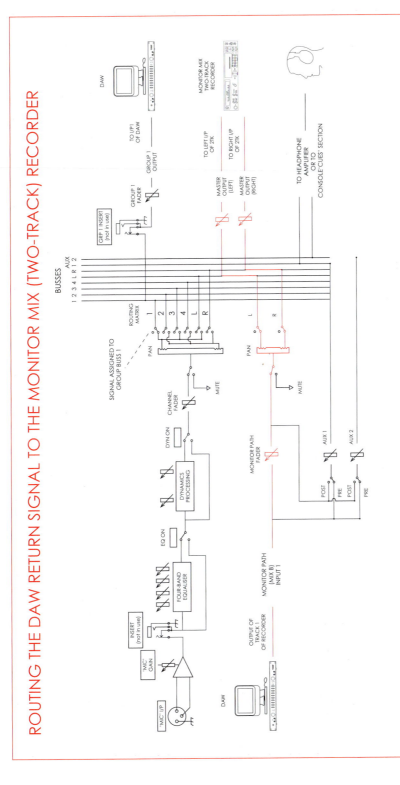

Continuing from the previous diagram, the output of the DAW is fed onto the monitor or Mix B path of the same console. The signal subsequently travels through a fader and a pan pot after which it reaches a second, smaller routing matrix where it is assigned to the master output buss. This main output is then fed onto the two-track device responsible for the recording of the monitor or reference mix.

THE CUE MIX FEED

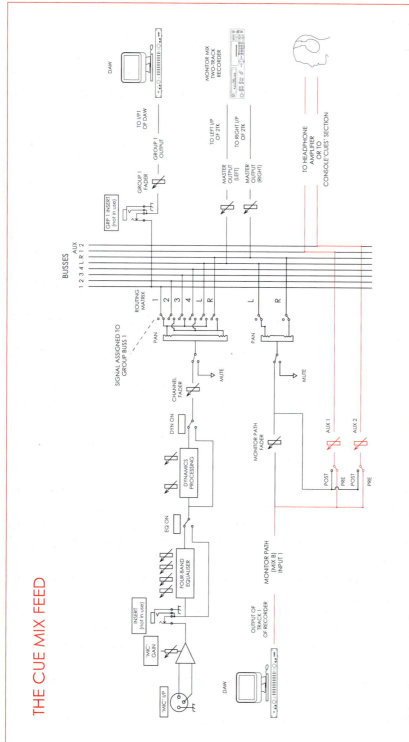

This diagram illustrates the use of two mono auxiliary busses for the creation of a cue mix. The signal is sourced from the monitor or Mix B path, **pre-fader** so the engineer is free to change the level balance and panning of the reference mix with no consequence to what the musicians hear through their headphones.

PATCHBAYS

Patchbays allow for the easy interconnection between the various devices found in a music production set-up. The flexibility they provide may help increase the efficiency of the recording process significantly, where the sometimes demanding and time-consuming equipment patching procedure is avoided or simplified.

Patchbays vary in kind (as audio connectors) and in size, spanning from models with a few domestic 'jack' patch points to large 'B-gauge' or bantam professional models. The following are some connectivity options found in patchbays used alongside in-line consoles:

- Microphone lines – Where the 'live room' microphone cables terminate, N.B. the term may appear confusing, as 'lines' is used here in place of 'leads' or 'cables' (not implying line level).
- Microphone inputs – Access point to the microphone level channel input of the console.
- Multitrack returns – Where the cables originating from the multitrack recorder output terminate.
- Channel line inputs – Access point to the line level channel inputs of the console.
- ('Off-tape') monitor input – Access point to the monitor path.
- Channel insert sends – Where signals may be sourced from for serial processing, e.g. dynamics.
- Channel insert returns – Access to a point where signals may be replaced with a processed version, e.g. equalised, compressed, etc.
- Group outputs – Where group buss signals may be sourced from, e.g. for the feeding of effects processors in mixdown.
- Multitrack sends / Off-buss monitor path input – Access point to the multitrack recorder inputs and to monitor paths (with a 'send' signal).
- Auxiliary or effects sends – Where signals may be sourced from for parallel processing, e.g. reverb, delay, etc.
- Auxiliary or effects returns – Access to especial console inputs, commonly fed by the output of effects processing units.
- Main outputs – Where the main stereo (mix) buss signal may be sourced from.
- Two-track input – Access to the two-track master mixdown recorder.

The following illustration is an example of a patchbay wired for a studio containing:

- An in-line console offering:
  - Twenty-four I/O modules
  - Eight mono auxiliary busses
  - Four stereo auxiliary returns
  - One main stereo output buss
  - Two 'external' stereo control room monitoring inputs
- Three external effects processors
- One two-track recorder (for the recording of mixes)
- A DAW used as a multitrack recorder.

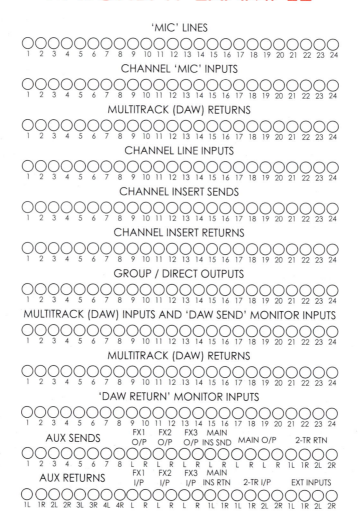

PATCHBAY EXAMPLE

'MIC' LINES

CHANNEL 'MIC' INPUTS

MULTITRACK (DAW) RETURNS

CHANNEL LINE INPUTS

CHANNEL INSERT SENDS

CHANNEL INSERT RETURNS

GROUP / DIRECT OUTPUTS

MULTITRACK (DAW) INPUTS AND 'DAW SEND' MONITOR INPUTS

MULTITRACK (DAW) RETURNS

'DAW RETURN' MONITOR INPUTS

AUX SENDS

AUX RETURNS

Patchbay Normalling

In the previous patchbay example, the following rows would commonly be 'normalled':

'Mic' Lines – 'Mic' Inputs
A fully-normalled connection allowing operators to reroute microphone signals originating from a live room ('mic' lines) to non-corresponding channel 'mic' inputs, e.g. the signal from the microphone connected into the first patch point in the live room wall box ('mic' line 1) can be rerouted ('cross-patched') onto Channel 25 (as opposed to the default Channel 1). This facility is useful when a console has a non-functional I/O module or 'mic' preamplifier or when the wall box in the live room has a bad connection. Full-normalling is used here to avoid the splitting of a microphone signal between two channels and/or the connection of two microphones and one 'mic' preamplifier in a circuit.

Multitrack Returns – Channel Line Inputs
A half-normalled connection found in true in-line consoles allowing operators to use a copy of the multitrack return signal without breaking the feed to the channel path's line input. A patching into the 'Channel Line Input' row will otherwise break the aforementioned normalling, i.e. the patched signal will replace the multitrack return signal feeding the channel path's line input by default.

Channel Insert Sends – Channel Insert Returns
A common half-normalled connection where operators can use a copy of the insert send signal, e.g. for parallel processing, while the patching into insert return will replace the signal in the original path.

Group / Direct Outputs – MTK Sends / 'MTK Send' Monitor Inputs
A half-normalled connection found in true in-line consoles where the group buss or the 'direct' output' signal ('multitrack send' signal) can be fed onto the monitor path. A patching into the upper row will provide operators with a copy of the group buss or the 'direct out' signal, which may be used for a variety of purposes such as cross-patching to a non-corresponding track, e.g. Group 1 feeding Track 2, or the routing of group

Patchbay Normalling (continued)

signals to effects processors during mixdown (commonly used when all auxiliary busses are already in use).

Multitrack (DAW) Returns – 'MTK Return' Monitor Inputs
Another half-normalled connection found in true in-line consoles where the multitrack recorder also feeds the monitor path via its own dedicated input. A patching into the upper row will provide operators with a copy of the multitrack return signal, while the patching into the lower row will force a signal onto the monitor path's default input.

Main Output – Two Track Input
A half-normalled connection feeding the main output buss signal to the two-track recorder. The patching onto the upper row provides users with a copy of the buss signal while the patching into the lower row replaces the signal feeding the two-track recorder.

Two-Track Return – External Input
A half-normalled connection allowing for the audition or confidence monitoring of the two-track recording in the control room. Again the upper row provides a copy of a signal (in this case the two-track recorder return) while the bottom row represents a break in the normalling allowing for the quick routing of signals to the control room monitoring section (useful for the playing of devices such as mp3 players, etc.)

The following pages describe the process of patchbay normalling graphically.

HALF-NORMALLED CONNECTION

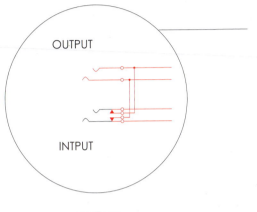

If no connector is inserted, the signal found in the top row is fed onto the bottom one.

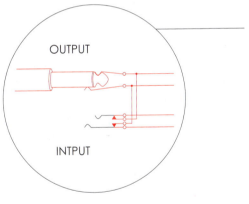

A connector patched into the top row will carry the signal found there.

The feed to the lower row will not be interrupted.

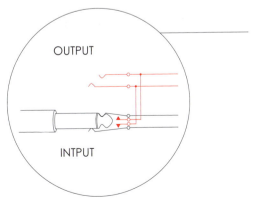

A connector patched into the bottom row will interrupt the feed from the top.

FULLY-NORMALLED CONNECTION

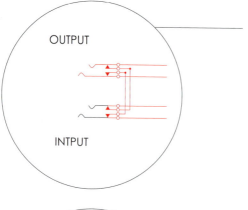

OUTPUT

INTPUT

If no connector is inserted, the signal found in the top row is fed onto the bottom one.

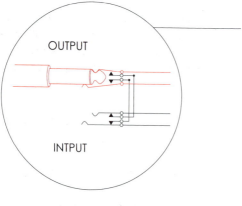

OUTPUT

INTPUT

A connector patched into the top row will interrupt the feed to the bottom.

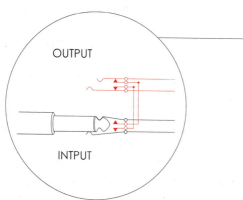

OUTPUT

INTPUT

A connector patched into the bottom row will interrupt the feed from the top.

EFFECTS PROCESSORS

A number of effects processing devices are utilised in recording and mixing sessions. These may be grouped according to their principles of operation and split into the following categories:

- Time-Delay-Based Effects
- Amplitude-Based Effects
- Filter-Based Effects
- Waveform Distortion Effects
- Other Effects.

The following sections describe common effects processors used in music production.

Time-Delay-Based Effects

Reverberation

Artificial reverberation devices are commonly used to add depth or dimension to signals. Some digital reverb presets are designed to simulate physical spaces, e.g. halls, chambers, rooms, etc., while others are emulations of electro-mechanical devices, e.g. spring and plate reverbs. It is common for reverb units to offer the following user-changeable parameters:

- Size or decay – The overall dimensions of the space or of the electro-mechanical device being simulated, i.e. the length of reverberation.
- Pre-delay – The time separating the 'dry' or direct sound from its first early reflection.
- (HF) Damping – A level attenuator affecting the high-frequency content of a reverberation effect.
- (LF) Damping – Similar to HF damping, LF damping allows for the control over the level of low-frequency content.
- Crossover – The point where the spectrum is divided into high and low frequencies, this parameter affects high and low-frequency damping.
- Density – The spacing between echoes and, therefore, the total number of reflections that fill up the reverberation decay time.
- Diffusion – The overall geometry of the reflection cluster, i.e. how unevenly the spacing between reflections develops over time.
- Mix – The relationship (ratio in percentage) between 'wet' and 'dry' sounds.

Practical Use of Reverberation Parameters
- Pre-delay – Longer pre-delays are employed to simulate environments with widely spaced boundaries (walls) and may help improve the clarity of focal elements, e.g. main vocals, solo electric guitar, etc.
- (HF) Damping – 'Darker' sounding artificial reverberation is generally perceived as more realistic.
- (LF) Damping – The attenuation of low-frequency content commonly helps clear 'muddy' or 'cluttered' reverberant sounds.
- Density – Dense reverberation is usually more suitable for percussive sources, e.g. drums (less individual sounding echoes), while sparse effects may suit vocals and other less transient sounds.
- Diffusion – Diffusion plays a very important role with longer decay presets, making the latter less artificial sounding.

NB Diffusion and density changes are easier to perceive when percussive sound sources are used.

Echo / Delay

Although similar to reverberation, echo or delay effects are commonly used in a more creative or less realistic fashion, due to the discrete or discernible nature of the reflections they produce. The following are common echo / delay parameters:

- Time or delay – The gap between echoes or copies of the 'dry' signal.
- Feedback – The total number of echoes. NB negative feedback implies the echoes are opposite in polarity in relation to the 'dry' signal.
- (HF) Damping – Same as in reverberation.
- (LF) Damping – Same as in reverberation.
- Crossover – Same as in reverberation.
- Mix – Same as in reverberation.

Modulated Delay Effects

Modulated delay units generate dynamically changing effects

Modulated Delay Effects (continued)

including 'flanging', 'chorusing', 'vibrato', 'harmonising', etc. This is commonly achieved through the use of low-frequency oscillators (LFOs) manipulating delay times. The following are basic parameters commonly found in modulated delay, flanger, chorus and vibrato devices:

- Delay – The base delay value altered by the modulation.
- Depth – The amount of delay modulation, i.e. the amplitude of the LFO wave.
- Rate or speed – The speed at which the delay time is changed, i.e. the frequency of the LFO.
- Waveform – The shape of the LFO wave, e.g. sine, square, triangle, etc.
- Resonance / Feedback / Regeneration – An internal feedback control making modulated delay effects more pronounced.
- Manual – The original centre frequency of modulation, i.e. the 'ringing' frequency.

Flanger and Chorus Modulated Delay Times

The following approximate modulated delay times may be used for the generation of flanger and chorus-like effects:

- Flanger – From 1 ms to 15 milliseconds (approximately)
- Chorus – From 15 ms to 30 milliseconds (approximately)

Natural Echo / Reverberation

Some commercial studios incorporate purpose-built natural reverberation chambers, containing amplifiers / loudspeakers and microphones for the production and pick-up of reverberation. Such chambers may contain different materials with variable acoustic properties, allowing for the manipulation of reverb characteristics.

Recordists on a budget may explore the use of reflective environments such as stairwells, garages, bathrooms, etc. for the purpose of adding natural reverberation to recordings. It is nevertheless important to note that small rooms may add unflattering colouration to recordings due to low-mid frequency resonances (which may clash with the key of the song performed and compromise clarity).

DELAY AND REVERB UNITS

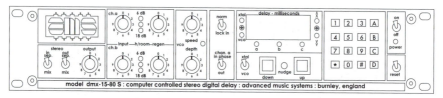

AMS DMX-15 (DELAY)

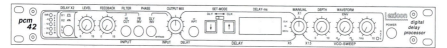

LEXICON PCM42 (DELAY)

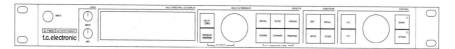

T.C. ELECTRONC D2 (DELAY)

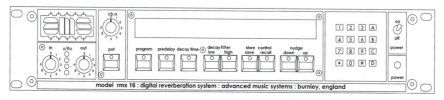

AMS RMX-16 (REVERB)

LEXICON PCM96 (REVERB)

BRICASTI M7 (REVERB)

DELAY PLUGINS

PRO TOOLS' AIR DYNAMIC DELAY

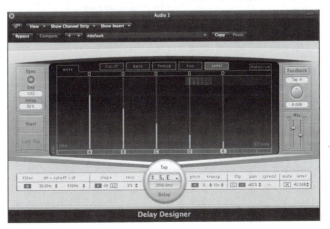

LOGIC'S DELAY DESIGNER

REVERB PLUGINS

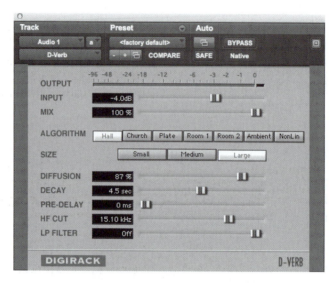

PRO TOOLS' D-VERB

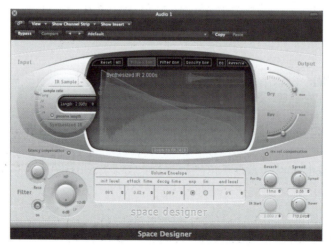

LOGIC'S SPACE DESIGNER

Amplitude-Based Effects

Tremolo

Tremolo units alter the level of signals in a cyclical manner, generally producing a 'stutter-like' effect. Common tremolo parameters include:

- Depth – The amount of amplitude modulation, i.e. the intensity of the effect, dictated by the level of the modulating LFO wave.
- Rate or speed – The speed at which amplitude is modulated or altered, i.e. the frequency of the LFO.
- Waveform – The shape of the LFO wave, e.g. sine, square, triangle, etc.

Ring Modulation

The principle of operation of ring modulators is similar to that of tremolo devices, although a high-frequency oscillator (or an upper range LFO) is used. Ring modulators generally produce metallic or robotic sounding results. The parameters found in ring modulators are similar to those of tremolo units.

Auto-Panning

Auto-panners move signals dynamically across the stereo field. Auto-panning device parameters include:

- Depth – The intensity of the panning effect.
- Rate or speed – The speed of panning.
- Waveform – The shape of the LFO wave, e.g. sine, square, triangle, etc.
- Width – How much of the stereo field is covered by the effect.

Filter-Based Effects

Phaser

Although sometimes described as another by-product of delay modulation, phasing is effectively achieved through the use of sweeping notch filters, as an altering of delay time would incur in pitch shifting (traditionally not a characteristic of the so-called 'phaser' effect).

Phasers also employ LFOs (used for the modulation of filter cut-off frequency) and present the following user-controllable parameters:

- Depth/Sweep – How far the notch filters will be swept, i.e. the amplitude of the LFO.

- Rate or speed – The speed at which the notch frequencies are swept, i.e. the frequency of the LFO.
- Waveform – The shape of the LFO wave, e.g. sine, square, triangle, etc.
- Feedback/Resonance – An internal feedback control, i.e. the output of the filters is fed back into their input.

Wah-Wah

The effect known as 'wah-wah' is accomplished through the use of a low-pass filter with a resonant peak. The filter's cut-off frequency may be modulated 'manually', e.g. by a foot pedal, or automatically by an LFO.

'Auto-wah' effects processors may also utilize envelope following circuitry as a source of modulation, i.e. the cut-off frequency is made directly proportional to the amplitude of the input signal.

Vocoder

In the vocoder effect, the characteristics of one signal, e.g. vocals, are superimposed onto those of another, unrelated 'carrier' signal, e.g. a synthesiser.

Waveform Distortion Effects

Overdrive / Distortion / Fuzz

The original idea behind the overdrive effect was to provide a gain boost to instrument level signals, e.g. electric guitars, on the way to the amplifier. This gain boost was set so the input stage of the amp would be overdriven, resulting in distortion. Overdrive effects processors, e.g. guitar pedals, were devised to generate their own internal distortion.

Distortion and fuzz effects offer progressively more extreme manipulation of waveforms, where the source signal is clipped more radically in either a symmetrical or non-symmetrical fashion. These effects were originally devised to allow instruments to produce harmonically rich or very aggressive sounds at low(er) amplifier levels.

Other Effects

Other effects are found in use during recording / mixing sessions. These include aural exciters, Leslie cabinets, talk-boxes, pitch correction applications, etc. Some of them correspond to a combination of different manipulators into a complex effect, e.g. 'exciters' and Leslie, while others are unique in the way they operate.

The Use of Effects in Recording / 'Printed' vs 'Monitor' Effects

Most recordists seem comfortable employing signal processing for the creation of rough (monitor) mixes, 'cue' mixes and other procedures that affect multitrack return signals, i.e. processes that do not alter what is being captured. The picture becomes quite different in regards to the use of effects in the recording signal chain, modifying what is being sent to the multitrack recorder, e.g. DAW. Here we seem to encounter two distinct schools of thought: the first is comprised of individuals who attempt to capture all sound sources free of effects, aiming to obtain a greater degree of flexibility during mixdown, while the second school consists of those who commonly record their signals 'wet', committing to desirable sounds as early as possible.

There are obvious advantages to both approaches and recordists should be ready to work following either of them. For safety, both the 'dry' and 'wet' versions of a signal may be recorded simultaneously.

Plug-ins

Most classic hardware effects processors have been emulated and now are available in software form. This is not to say that the sonic qualities of the two counterparts are identical (and in fact this may not be relevant), although in some cases the similarity may seem surprising.

The effects parameters described here still apply, and in some cases may be expanded, in 'virtual' effects processing.

Multi-Effects Units

It is common for digital units to offer multiple types of effects and presets. Such devices may also:

- Incorporate more than one processing engine, i.e. they may allow users to choose two or more effects to be used simultaneously.
- Allow the user to change the configuration (mono/stereo) and the order of processing, e.g. delay – reverb or reverb – delay, etc.

MULTI-EFFECTS UNITS

YAMAHA SPX2000

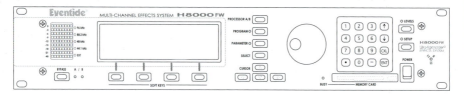

EVENTIDE H8000FW

LEXICON 960L

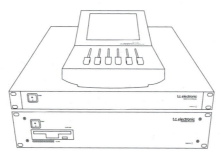

TC ELECTRONIC SYSTEM 6000

MULTITRACK RECORDERS

Multitrack recorders (MTR) allow for operators to capture numerous sound sources simultaneously, while storing them in isolation, i.e. several instruments may be recorded at the same time onto their own unique track. This provides great control over productions during the subsequent mixdown stage, e.g. a unique level balance between instruments may be generated, individual sources may be muted, etc.

Analogue tape machines were the first 'sound on sound' multitrack recorders utilised in commercial studios, evolving from two to three, four, eight, sixteen, twenty-four and thirty-two track capabilities. With the advent of digital audio, different devices took the place of the analogue tape recorder. Some of them are standalone, e.g. ADAT, RADAR, etc., while some, commonly referred to as DAWs, represent a combination of hardware, e.g. computer/audio interface, and software, e.g. Logic Pro, Pro Tools, etc.

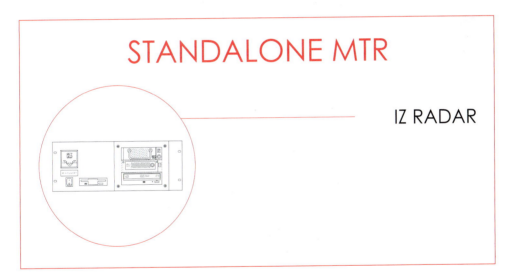

STANDALONE MTR

IZ RADAR

TWO-TRACK RECORDERS

Two-track recorders are commonly used to capture 'reference' or monitor mixes during recording sessions. Such devices are usually fed with the output signal from the console's main stereo buss and are commonly kept running at all times during production, so as to capture any spontaneous ideas, dialogue between musicians, etc.

It is important to note that many recordists prefer to record the two-track mix onto the multitrack recorder itself.

TWO-TRACK RECORDER

ALESIS MASTERLINK

THE DIGITAL AUDIO WORKSTATION (DAW)

'Digital Audio Workstation' is the collective name given to the components that make it possible for analogue audio to be converted and stored digitally. The number of separate elements that sum up to a DAW may vary from one (a standalone recorder) to many more.

The following are examples of possible DAW configurations:
- Audio Interface > Computer (running Logic, Pro Tools, etc.)
- Standalone A/D (and D/A) Converter > PCI Card > Computer
- Standalone A/D (and D/A) Converter > Audio Interface > Computer.

The A/D – D/A Converter

The A/D – D/A converter is an extremely important element of the digital audio recording chain and may be found in standalone form or as part of an audio interface. Converters commonly offer balanced line level, XLR or 'jack' connections on the analogue end and can operate with one or multiple digital audio formats, e.g. AES/EBU, SPDIF, ADAT Optical, etc.

The output of standalone A/D converters is normally connected to the digital input of an audio interface, which is in turn linked to a computer via Firewire or USB connections. Some specialised devices work in tandem with computer PCI cards, bypassing the need for an external interface. The connection between such converters and their corresponding computer cards is commonly made through D-Sub links.

A/D – D/A CONVERTERS

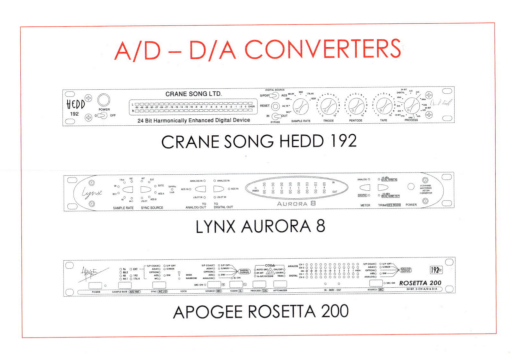

CRANE SONG HEDD 192

LYNX AURORA 8

APOGEE ROSETTA 200

The Audio Interface / Sound Card

The audio interface may be described as the computer's digital audio input device. It may incorporate the functions of an A/D – D/A converter or simply accept digital signals, routing them to a computer through a PCI card or via USB or Firewire links.

AUDIO INTERFACES

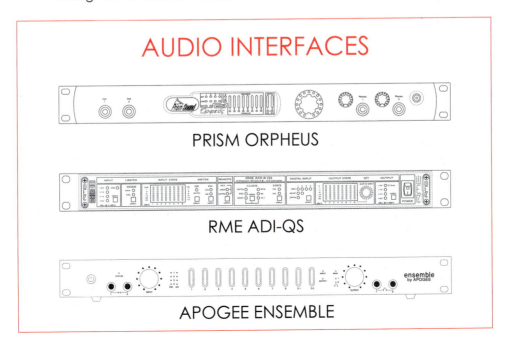

PRISM ORPHEUS

RME ADI-QS

APOGEE ENSEMBLE

The Computer / Software Application(s)

A number of software applications are currently utilised for recording purposes, e.g. Cubase, Digital Performer, Logic Pro, Pro Tools, Record, Reaper, etc. With the passage of time, these are becoming increasingly similar and at present the choice of program appears to be a simple matter of taste, i.e. a good recordist should be able to work using any of the popular DAW software applications running on either a PC or Macintosh OS-based computers.

NB Other devices may be found as part of DAW systems, e.g. word clock generators, control surfaces, etc.

Word clock

Digital devices require a common clocking signal when operating as elements of the same signal chain. Such signal is referred to as word clock and it may be described as a series of electrical pulses occurring at regular intervals (matching the effective sampling rate).

The source of word clock must be extremely stable and as a rule of thumb, recordists should always set the clock output of the device of highest manufacturing quality in the chain as the 'master' (which may in some cases simply equate to the most expensive device acting as the timing reference).

Jitter

Inconsistencies in digital clocking may lead to a phenomenon known as 'jitter', commonly detected as the audio signal appears distorted or as 'clicking' sounds are produced by the system.
The most effective way to avoid jitter is to ensure that a single, exceedingly stable device is used as the word clock master and all other elements are set as 'slaves'.

Control Surfaces

Control surfaces may improve the DAW workflow significantly,

Control Surfaces (continued)
allowing recordists to operate the system with more efficiency.

Such devices may incorporate faders, switches, rotary encoders, etc. and may be very useful in recording sessions that do not incorporate an analogue console ('in the box'), e.g. for the setting of 'cue' and stereo monitor mixes, etc.

THE MONITORING SYSTEM

The monitoring system, alongside the input to the signal chain, i.e. the transducers, should be given the highest priority in the audio path hierarchy. Without a fair and somewhat transparent evaluation of the material being recorded, production teams may work on the tracking stage of a project only to discover subsequently that the content gathered is less than ideal and does not correspond to expectations.

Monitoring systems may incorporate one or many of the following elements:

Power Amplifiers

Power amplifiers raise line level signals to loudspeaker level. This significant increase in voltage levels may be achieved in various manners and 'amps' can differ from each other considerably. Many of the factors that contribute to the aforementioned differences, e.g. topology, are beyond the scope of this book, which focuses on providing recordists with basic guidance.

The quest for clean, non-distorted sound reproduction should start with an investigation into amplifier and loudspeaker power. Equipment 'spec sheets' commonly display a number of different figures of varying relevance. One of these, 'continuous power rating' appears as the most important and realistic, as it describes the ability of a device to continuously handle 'real-world' program material. As a rule of thumb, the amplifier should present no less than twice (and ideally four times) the IEC continuous power rating of a connected loudspeaker with matching nominal impedance, e.g. a power amp with a continuous IEC power rating of 400 watts at 8 ohms should be used to feed a loudspeaker rated at 100 watts at 8 ohms (or at the very least a 200-watt power amp should be used). This ensures the amplifier will operate far below stress levels and with plenty of headroom.

Another important specification, known as 'slew rate' describes the ability of an amplifier to follow (react to) input transients and it is expressed in volts per microsecond. The slew rate dictates how high the input sensitivity of an amplifier (level) may be set to until it cannot reproduce transients faithfully. A very generic way to delineate suitable slew rates is:

Rate > 10 V/µs   For amplifiers rated under 100 watts per channel
Rate > 30 V/µs   For amplifiers rated over 200 watts per channel

Current high-quality power amps have slew rates that fall well above the described minimum range without presenting objectionable distortion or suffering from interference.

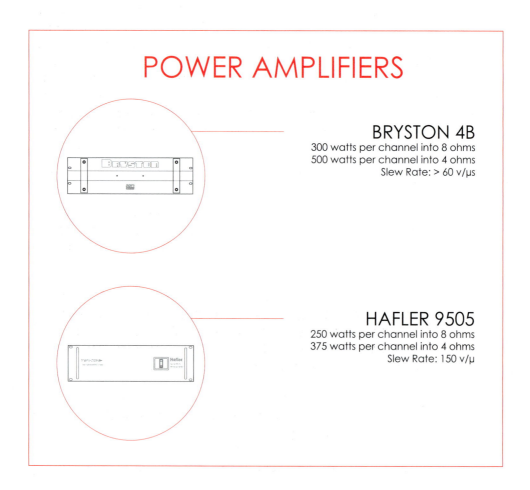

POWER AMPLIFIERS

BRYSTON 4B
300 watts per channel into 8 ohms
500 watts per channel into 4 ohms
Slew Rate: > 60 v/µs

HAFLER 9505
250 watts per channel into 8 ohms
375 watts per channel into 4 ohms
Slew Rate: 150 v/µ

Loudspeakers
Loudspeakers are commonly the last and possibly the most important elements in the signal chain. Still, it is not uncommon for their selection to be heavily influenced by subjective criteria, where choice is based primarily

on 'character' and not on fidelity (in a similar fashion to microphones). The topic of studio monitoring is a heavily debated one and it is approached with extreme caution by some, while others view it as a simple matter of taste. Despite the argument, it is nonetheless clear that different speaker models may be used effectively for music production as long as the technicians that employ them understand their behaviour.

A case may be made supporting the use of 'flattering' or 'colouring' monitors, considering that transparency may be virtually unobtainable for recordists working with restricted budgets and that ultimately most consumers will listen to music through uneven systems. With that in mind, it is not unusual for engineers to choose to work with loudspeakers that reproduce signals with a familiar bias or 'colouration', i.e. devices that superimpose a desirable uneven frequency response onto that of the program material. On the other hand, although it is possible for recordists to achieve good results with basic budget monitors, in the long run it seems advisable for them to seek for the 'flattest', most 'truthful' high-quality speakers, as such devices should lead to consistent results that translate well to a greater range of systems, e.g. from portable mp3 players to large PAs.

Specification Sheets

Specification or 'spec' sheets may help engineers wishing to choose their monitors using more than just personal taste. A few technical aspects appear as most important and should be taken into consideration during the selection process, including:

- Free-Field Frequency Response – This figure provides an overall view of a loudspeaker's frequency response within an acceptable level of deviance (ideally within 1 dB and always below 3 dB), e.g. 50 Hz to 20 kHz, ± 1 dB. Such information is commonly presented in graphic form.

- Cut-Off Frequency (Lower and Higher) – The extreme ends of a loudspeaker's frequency response, where reproduction is at a negative 3 dB deviation from unity gain, e.g. – 3 dB at 47 Hz and 22 kHz. This measurement may not be relevant in cases where the free-field frequency response of a loudspeaker is stated.

Maximum SPL – A general indication of the maximum output level for a frequency range at a given distance, which should be accompanied by a THD figure, e.g. maximum sound

Specification Sheets (continued)

pressure level at 1 metre (100 Hz to 10 kHz) = 110 dBSPL (at 3 percent THD).

Total Harmonic Distortion (THD)

Total harmonic distortion is a measurement describing the non-linearity of devices. THD is calculated through a comparison between the harmonic content at the input and output stages of a given piece of equipment. Total harmonic distortion may be expressed as a number or as a percentage and, as a rule, low THD values imply high quality in manufacturing. THD readings of 3 percent and higher should be considered unacceptable.

Driver Types

Loudspeaker drivers are transducers, converting electron flow into acoustic energy. This conversion is made possible by processes that are identical (although opposite in direction) to those employed by microphones. The following are the most common types of drivers found in loudspeaker enclosures:

- Dynamic – Moving Coil
  The most popular design, moving coil drivers are very resilient and therefore able to reproduce sound at high levels (extreme excursions). They are suitable for operation at different frequency ranges and are used as sub-woofers, woofers and tweeters (varying in size).

- Dynamic – Ribbon
  Ribbon drivers are fragile and suitable for high-mid / high-frequency, (relatively) low sound pressure level excursions only (tweeters).

- Electrostatic
  Electrostatic drivers reproduce sound with minimal distortion

Driver Types (continued)

and excellent timing, i.e. transient response, although at (relative) low output levels. Such drivers are not commonly employed for the reproduction of low-frequency sound content due to physical restrictions.

- Piezo
  Piezo drivers are used for the generation of high / very high-frequency content by tweeters and super-tweeters (limited excursions). This type of transducer is found in some types of headphones, etc.

Cabinet (Enclosure) Design

Enclosures were designed to decouple the front and the rear of loudspeaker drivers, as the two are opposite in polarity, i.e. if drivers were to be simply suspended in mid-air, most low-end acoustic energy would be 'cancelled' as the back of a driver would 'suck' the air displaced at the front and vice-versa. An ideal way to avoid this 'coupling' would be to mount drivers on very large panels or 'baffles' (ideally infinite in size), which would ensure the separation between their two poles, with no side effects (except for the loss of 50 percent of all energy created). As this is not practical, a few alternate ways to improve the efficiency of driver systems were devised.

- Sealed Enclosure
  A sealed enclosure is a box with no openings or ports (except for very small air 'leaks'). Such a box guarantees the separation between the front and the rear of the transducer, although a significant amount of energy is wasted. The air inside a sealed enclosure acts as a spring, opposing the movement of the cone and, as a consequence, closed cabinets must be relatively larger in size to allow for low-frequency reproduction.

  Some individuals refer to mounted studio speakers as 'infinite baffles', although such devices' behaviour is closer to that of sealed enclosures (without diffraction issues) as the air behind mounted drivers will still offer resistance to cone movement.

Cabinet Design (continued)

- Ported / Bass Reflex Enclosure
  Ported cabinets have vents or openings that make it possible for air (and therefore energy) to escape the enclosure. Such design allows for an extended bass response from smaller units (through resonance), although this commonly comes at the expense of added 'colouration'.

- Transmission Line
  Transmission Line enclosures may be thought of as vented boxes with an added inner acoustic 'labyrinth', through which sound has to travel before reaching the outside. This allows for the energy from the rear of the driver to be (more) coherent with that of the front, as it reaches the open air (due to the delay in travel).

- Auxiliary Bass Radiator (ABR)
  The Auxiliary Bass Radiator design is another attempt to utilise the energy from the back of loudspeaker drivers. In ABR cabinets, a 'dummy' (inactive) cone takes the place of a vent, reducing the stiffness of the 'spring' system of a closed enclosure, while using more of the rear energy (increasing efficiency).

Enclosure Shape

The shape of a cabinet affects how it will cope with internal pressure and how sound will diffract around it. For these reasons a few manufacturers have been moving towards rounded or curved designs.

Passive vs. Active Loudspeakers

Some loudspeakers, referred to as 'active', incorporate a built-in power amplifier. This implies they may be fed with line level signals, e.g. the main output of a mixing console. The advantages of such devices lie in the possibility of a perfect match between amplifier and drivers in the case of high-end manufacturers.

PASSIVE LOUDSPEAKERS

PMC DB1S+

Maximum SPL = 111dB SPL
Freq. Response = 50 Hz to 25 kHz (+/- N/A)
Design = Transmission line
Drivers = Moving coil

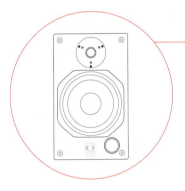

QUESTED H108

Maximum SPL = 110dB SPL
Freq. Response = 65 Hz to 20 kHz (+/- 2 dB)
Design = Vented enclosure
Drivers = Moving coil

YAMAHA NS-10M

Maximum SPL = N/A
Freq. Response = 85 Hz to 20 kHz (+/- 5 dB)
Design = Sealed enclosure
Drivers = Moving coil

ACTIVE LOUDSPEAKERS

GENELEC 1032A

Maximum SPL = 103dB SPL (< 1% THD)
Freq. Response = 42 Hz to 21 kHz (+/- 2.5 dB)
Design = Vented enclosure
Drivers = Moving coil

ADAM A7X

Maximum SPL = 114 dB SPL (< 1% THD)
Freq. Response = 42 Hz to 50 kHz (+/- 3 dB)
Design = Vented enclosure
Drivers = Coil (woofer) / Ribbon (tweeter)

KRK VXT8

Maximum SPL = 106.2dB SPL (3% THD)
Freq. Response = 50 Hz to 26 kHz (+/- 3dB)
Design = Vented enclosure
Drivers = Moving coil

Headphones

Headphones play an important dual role in music production, being used not only by performers but also by recordists on occasion. As far as engineering with headphones is concerned, the following are advantages of working with such devices:

- Headphones tend to provide recordists with a greater level of detail than 'middle of the road' / low-quality studio monitors.
- Headphones allow for work to be carried out under 'less than ideal' control room circumstances or even within the performance space, i.e. the 'live' room.
- It is very likely that current productions will be auditioned (by the general public) through headphones more extensively than through any other speaker-based monitoring system.

With that in mind, it is still important to also note the following disadvantages of working with headphones:

- The stereo image of headphones is considerably different to that of loudspeaker systems (unless crosstalk and delay are introduced).
- Engineers tend to work more 'conservatively' when mixing with headphones (lower relative levels, e.g. vocals, and narrower stereo image).

Ultimately, as artists and other production personnel will commonly expect to audition 'takes' or monitor mixes through conventional monitoring systems, headphones should be used temporarily and always alongside loudspeakers.

Types of Headphones
- Closed
  Example: Sony MDR-7506
  - Greatest isolation (in both directions)
  - Best for recordists working in the performing space
  - Best for musician's cues
  - May cause physical fatigue
- Semi-Open
  Example: AKG K141 MKII
  - A compromise between isolation and comfort
  - Isolation may be insufficient for cue mix use
- Open
  Example: Sennheiser HD800
  - Greatest comfort
  - Poor isolation

Monitoring Management Devices

Monitoring management devices aim to provide large format console, control room monitoring capabilities to individuals working in smaller, 'budget' DAW-based studios.   This type of equipment usually offers functions such as output level adjustment, input selection, output (monitor) selection, talkback, headphone outputs, etc.

MONITOR MANAGEMENT

DANGEROUS AUDIO D-BOX

MACKIE BIG KNOB

MUSIC PRODUCTION

RECORDING

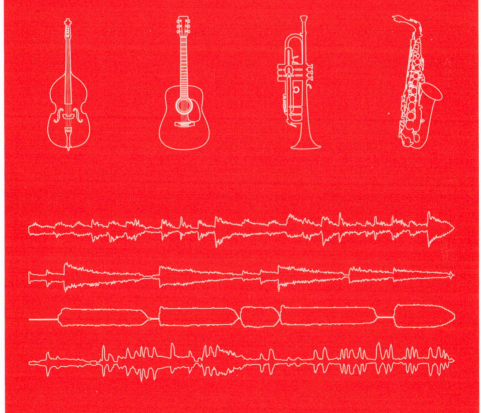

PRE-PRODUCTION

PREPARING TO RECORD

The music production process starts much before artists enter the recording space. The preparation that precedes sessions may ultimately dictate the potential for their success and it is vital for all involved to understand the importance of planning. With the confirmation of an upcoming project, producers and engineers should gather as much information as possible regarding the material to be recorded and the venues and tools that will be used for sessions.

RESEARCH

The initial familiarisation stage of pre-production should not only incorporate research on the work that the artists create (live shows, existing material, etc.) and the direction in which they want to take it, but also on their attitude towards and views on music and the industry. It is very important for the recording team to establish what the performers *think* their music sounds like as early as possible, as this should give an indication of what they expect to hear back from their recordings.

During planning, producers have the opportunity to discuss objectives and strategies with the performers, helping them 'map out' their desired session outcomes. This process may not be simple, as some individuals can find it difficult to discuss their influences and aspirations, e.g. inexperienced musicians may refuse to be compared to any pre-existing or contemporary acts and seem uncertain of what they want to achieve. Moreover, some artists might simply find the pre-production stage unnecessary and appear unwilling to co-operate. In such circumstances it is important for producers not to take such overly confident or detached attitude as final, as it could merely result from inexperience, lack of musical knowledge or insecurity. It is ultimately the job of the supporting team to ensure that artists feel comfortable in describing their vision, as this will help establish a frame of reference for production.

THE DEMO

The demo, i.e. an early, 'rough' recording of a composition, is a very powerful tool that should be used extensively during pre-production. Armed with it, producers and artists can examine the material that will be recorded and discuss possible ways to maximise its impact. In some circumstances, such recordings may also expose the weaknesses of performers, which can help the team focus the production on their strengths.

It is important to mention that the excessive referencing of 'demos' during the recording stage may be counter-productive and detrimental for sessions, as artists might get overly attached to their original performance (and its sound). In such cases, attempts to further improve a production may meet with constant opposition on the part of performers, who are anecdotally described as suffering from 'demo fever'.

Ideally, 'demos' should only be used as a reference during the pre-production stage, when they should be analysed and played alongside the artist's 'influences'. This may help establish what should realistically be achievable during production.

THE COMPOSITIONS

Although the knowledge of music theory may not be seen as a prerequisite for popular music production, an understanding of the basics of composition and arranging may help producers earn the respect and the trust of the artists they work with, which in turn may help facilitate the communication process and encourage the exchange of ideas. Ideally, during the pre-production stage artists should commit to a final form and arrangement for their songs, as this information will be used to guide the team throughout the recording process.

Form

The form of songs is commonly described in terms of sections, i.e. compositions are commonly broken down into:

- Introduction
- Verses
- Choruses (at times preceded by sections referred to as 'pre-chorus')
- Bridge(s)
- Interlude(s)
- Instrumental(s)
- Outro.

Such sections may be further divided into bars or measures (possibly arranged into blocks, etc.), allowing for information regarding song position to be communicated quickly and efficiently.

The following examples illustrate how the form of songs may be mapped out in a very basic pop producer or recordist-oriented format:

SOMETHING IN THE WAY

Words and Music: Kurt Cobain
Performer(s): Nirvana

INTRO	VERSE 1

CHORUS 1	VERSE 2

CHORUS 2

HERE COMES THE SUN

Words and Music: George Harrison
Performer(s): The Beatles

INTRO	CHORUS 1

VERSE 1	CHORUS 2

VERSE 2	CHORUS 3

BRIDGE

VERSE 3	CHORUS 4

CHORUS 5	OUTRO

Simplified charts outlining song form, such as the previous examples, or 'lyric sheets' (see page 289) may be sufficient for recordists aiming to run standard pop / rock recording sessions with a minimum degree of control over performance. Such charts may be further developed to display bars, beats, etc. for more complex sessions that require a greater command on the part of the producer, e.g. those involving classically trained musicians (where full scores are commonly used). Please refer to Appendix 2 for examples of more elaborate producer-oriented charts.

Arrangement / Instrumentation

Considering that Western popular music is mostly based on simple diatonic melodies and harmonies, lyrical content and instrumentation arguably hold the key to a song's impact. This does not imply that composers must always aim to write sophisticated lyrics and arrangements for their compositions, but that they should strive to find the most authentic 'feel' to deliver an engaging message distinctively.

Although it is important to note that lyrics play a vital part in the capturing of the listener's ears and imagination, without differences in texture, musical works based on preexisting, predictable patterns would most likely be perceived as repetitive and tedious by the general public, i.e. without contrast in instrumentation and dynamics, less inspired compositions could lack the power to hold the audience's attention for more than a few seconds. With this in mind, once the form of a song is established, the production team must examine all options regarding arrangement carefully.

The examples that follow illustrate how the arrangement of songs may be mapped out in a simple, pop producer or recordist-oriented format. Such charts may serve as an important visual aid outlining contrast and, possibly, the dynamic curve of a composition and may be used to help measure the progress of the recording stage, allowing for a quick scan of what is required for the completion of a given production.

SOMETHING IN THE WAY

Words and Music: Kurt Cobain
Performer(s): Nirvana

INTRO

A_GTR
VOX

VERSE 1

A_GTR
VOX

CHORUS 1

A_GTR	DRMS	VCL
VOX (x2)	BASS	

VERSE 2

A_GTR
VOX

CHORUS 2

A_GTR	DRMS	VCL
VOX (x2)	BASS	

HERE COMES THE SUN

Words and Music: George Harrison
Performer(s): The Beatles

INTRO

A_GTR		
		SYNTH

CHORUS 1

A_GTRs	VOX	DRMS	E_GTR
STRGS	BVs	BASS	

VERSE 1

A_GTRs	VOX	DRMS
STRGS	BVs	BASS

CHORUS 2

A_GTRs	VOX	DRMS	E_GTR
STRGS	BVs	BASS	

VERSE 2

A_GTRs	VOX	DRMS	
STRGS	BVs	BASS	SYNTH

CHORUS 3

A_GTRs	VOX	DRMS	E_GTR
STRGS	BVs	BASS	SYNTH

BRIDGE

A_GTRs	VOX	DRMS	E_GTR	CLAPS
STRGS	BVs	BASS	SYNTH	

VERSE 3

A_GTRs	VOX	DRMS	E_GTR
STRGS	BVs	BASS	SYNTH

CHORUS 4

A_GTRs	VOX	DRMS	E_GTR
STRGS	BVs	BASS	SYNTH

CHORUS 5

A_GTRs	VOX	DRMS	E_GTR
STRGS	BVs	BASS	SYNTH

OUTRO

A_GTRs	VOX	DRMS	E_GTR
STRGS	BVs	BASS	SYNTH

Arranging

Arranging is defined here as the act of selecting the instruments that will play at specific sections of a given recording. Some authors reserve this term for the reworking of an existing song for a different set of instruments than those for which it was originally composed.

Arrangement and Mixing

Since the introduction of multitrack recording, it has been possible for engineers to manipulate the arrangement of tracks during the mixdown stage of production, e.g. through mutes, solos, etc. It is not uncommon at present for performers to record brief static parts to be 'looped' over the entire length of songs, allowing producers to postpone important arrangement related decisions until the subsequent editing and mixing sessions. It is nonetheless important to remember that musicians will perform differently when they are made aware of their changing role within a given track, which may help bring to life an otherwise static musical composition, e.g. through the use of contrast / dynamics.

THE INSTRUMENTS

The knowledge of instrument characteristics, including part names (see Appendix 4), range, score attributes, etc. may lead to the crafting of efficient arrangements, facilitate communication in the studio and influence recording (and mixing) decisions, such as microphone choice, the use of processors such as compressors, pass filters, etc.

The following pages display information regarding some instruments that are commonly used in popular music production.

STRINGS
BASS

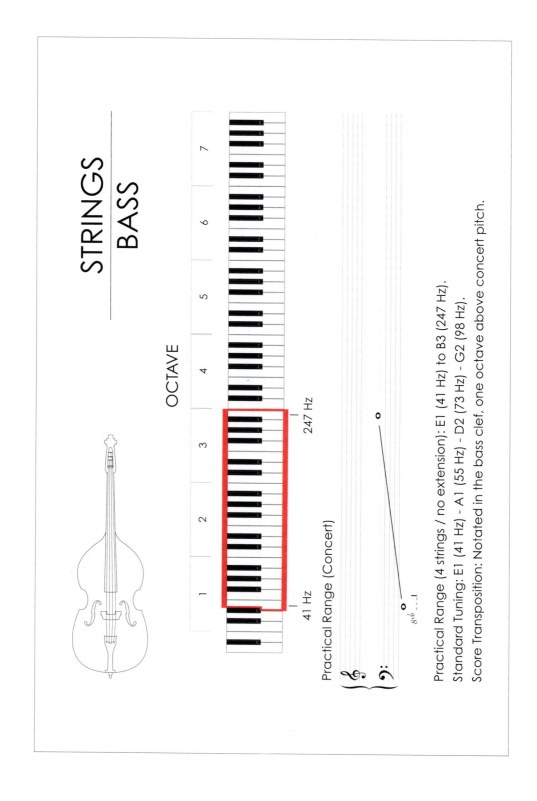

OCTAVE

Practical Range (Concert)

41 Hz

247 Hz

Practical Range (4 strings / no extension): E1 (41 Hz) to B3 (247 Hz).

Standard Tuning: E1 (41 Hz) - A1 (55 Hz) - D2 (73 Hz) - G2 (98 Hz).

Score Transposition: Notated in the bass clef, one octave above concert pitch.

STRINGS
CELLO

OCTAVE

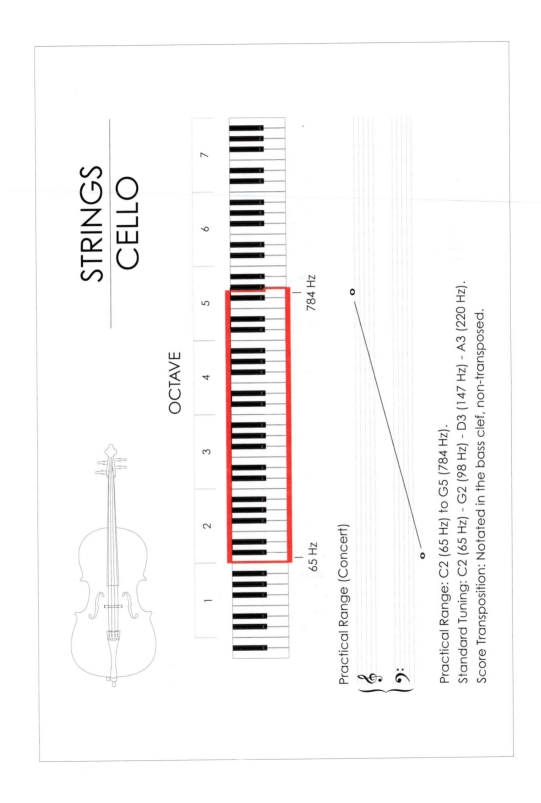

Practical Range (Concert)

Practical Range: C2 (65 Hz) to G5 (784 Hz).

Standard Tuning: C2 (65 Hz) - G2 (98 Hz) - D3 (147 Hz) - A3 (220 Hz).

Score Transposition: Notated in the bass clef, non-transposed.

STRINGS
GUITAR

OCTAVE

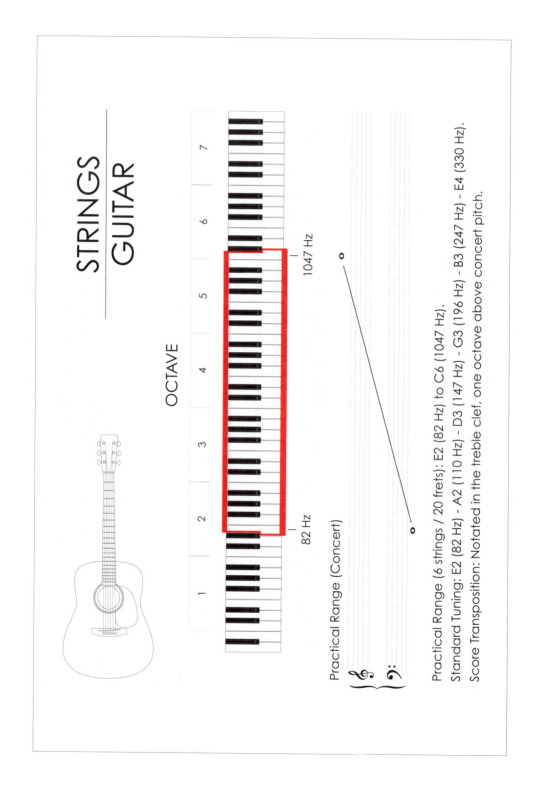

82 Hz

1047 Hz

Practical Range (Concert)

Practical Range (6 strings / 20 frets): E2 (82 Hz) to C6 (1047 Hz).
Standard Tuning: E2 (82 Hz) - A2 (110 Hz) - D3 (147 Hz) - G3 (196 Hz) - B3 (247 Hz) - E4 (330 Hz).
Score Transposition: Notated in the treble clef, one octave above concert pitch.

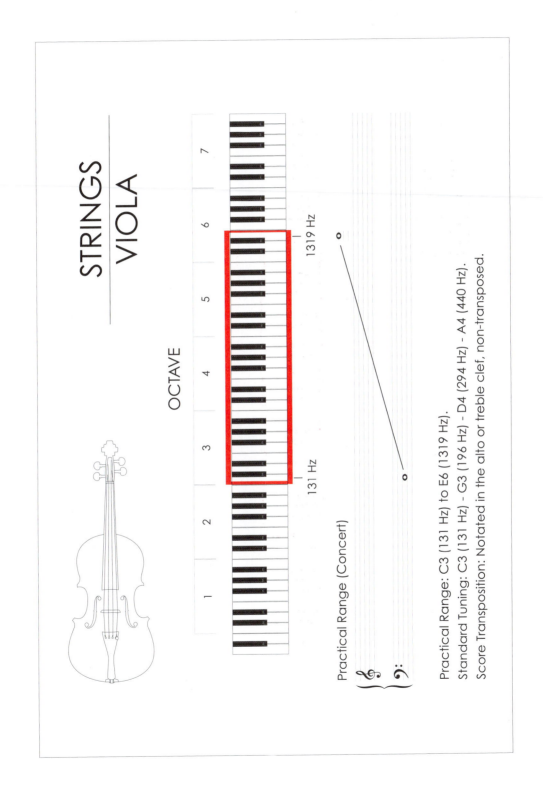

STRINGS
VIOLA

OCTAVE

1319 Hz

131 Hz

Practical Range (Concert)

Practical Range: C3 (131 Hz) to E6 (1319 Hz).
Standard Tuning: C3 (131 Hz) - G3 (196 Hz) - D4 (294 Hz) - A4 (440 Hz).
Score Transposition: Notated in the alto or treble clef, non-transposed.

STRINGS
VIOLIN

OCTAVE

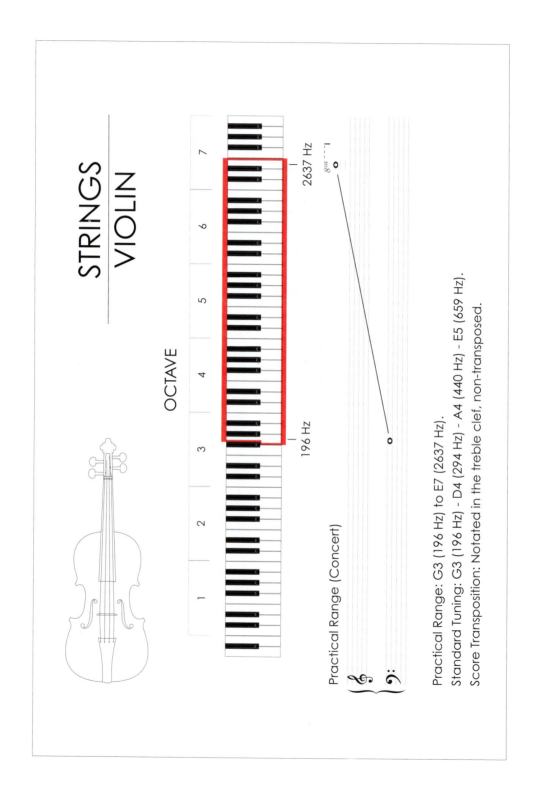

2637 Hz

196 Hz

Practical Range (Concert)

Practical Range: G3 (196 Hz) to E7 (2637 Hz).
Standard Tuning: G3 (196 Hz) - D4 (294 Hz) - A4 (440 Hz) - E5 (659 Hz).
Score Transposition: Notated in the treble clef, non-transposed.

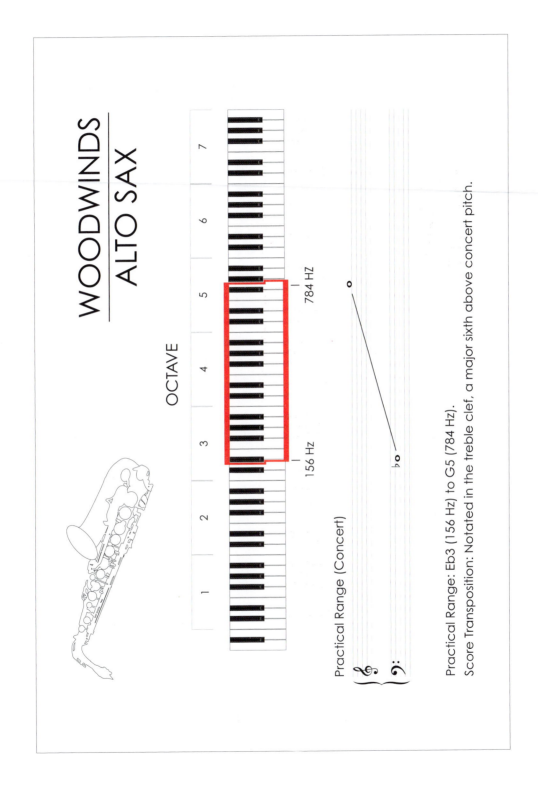

WOODWINDS
ALTO SAX

OCTAVE

784 Hz

156 Hz

Practical Range (Concert)

Practical Range: Eb3 (156 Hz) to G5 (784 Hz).

Score Transposition: Notated in the treble clef, a major sixth above concert pitch.

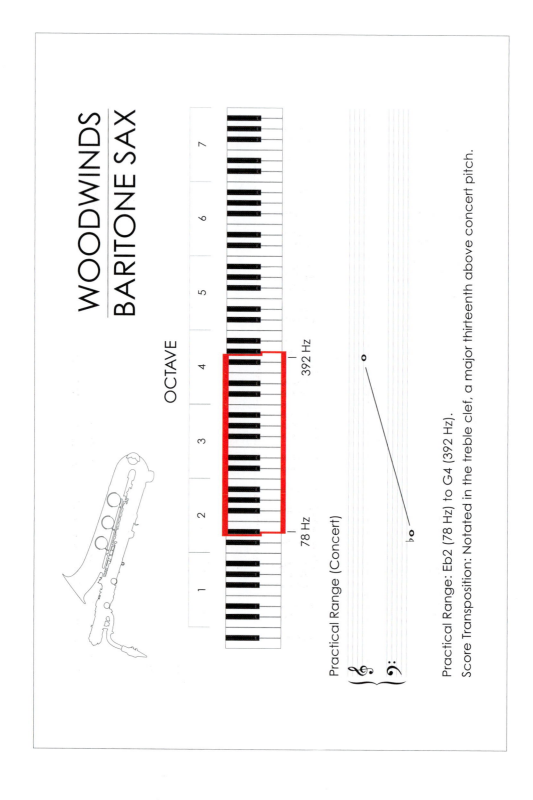

WOODWINDS
BARITONE SAX

OCTAVE

1　2　3　4　5　6　7

392 Hz

78 Hz

Practical Range (Concert)

Practical Range: Eb2 (78 Hz) to G4 (392 Hz).

Score Transposition: Notated in the treble clef, a major thirteenth above concert pitch.

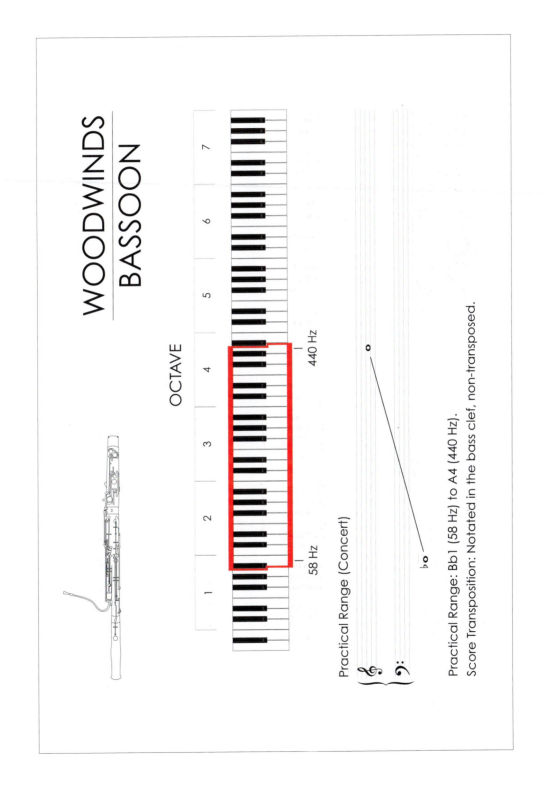

WOODWINDS
BASSOON

OCTAVE

440 Hz

58 Hz

Practical Range (Concert)

Practical Range: Bb1 (58 Hz) to A4 (440 Hz).
Score Transposition: Notated in the bass clef, non-transposed.

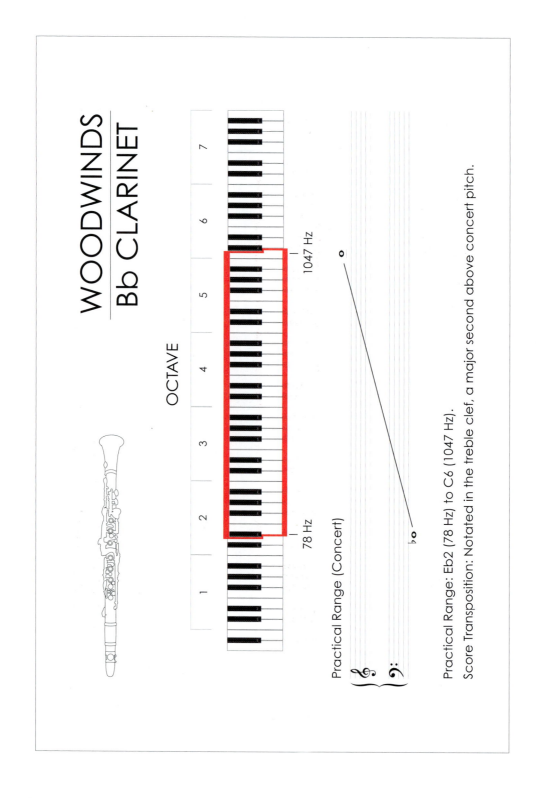

WOODWINDS
Bb CLARINET

OCTAVE

1 2 3 4 5 6 7

78 Hz

1047 Hz

Practical Range (Concert)

Practical Range: Eb2 (78 Hz) to C6 (1047 Hz).

Score Transposition: Notated in the treble clef, a major second above concert pitch.

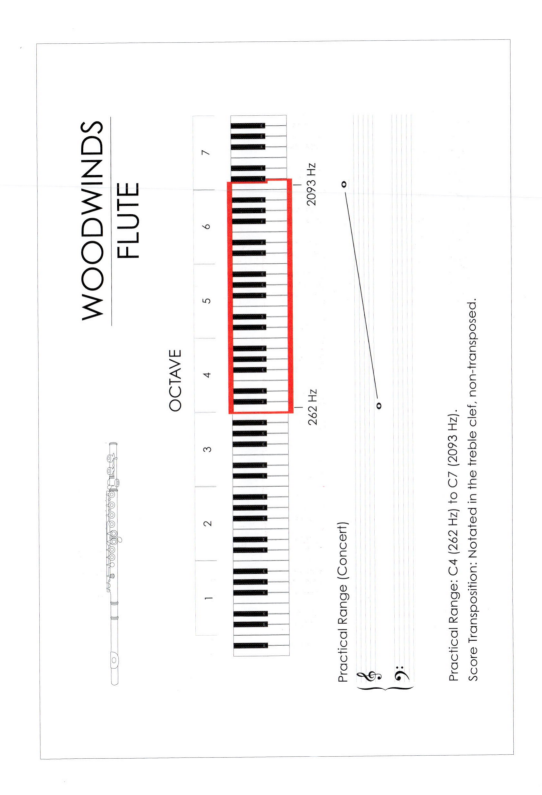

WOODWINDS
FLUTE

OCTAVE

2093 Hz

262 Hz

Practical Range (Concert)

Practical Range: C4 (262 Hz) to C7 (2093 Hz).

Score Transposition: Notated in the treble clef, non-transposed.

WOODWINDS
OBOE

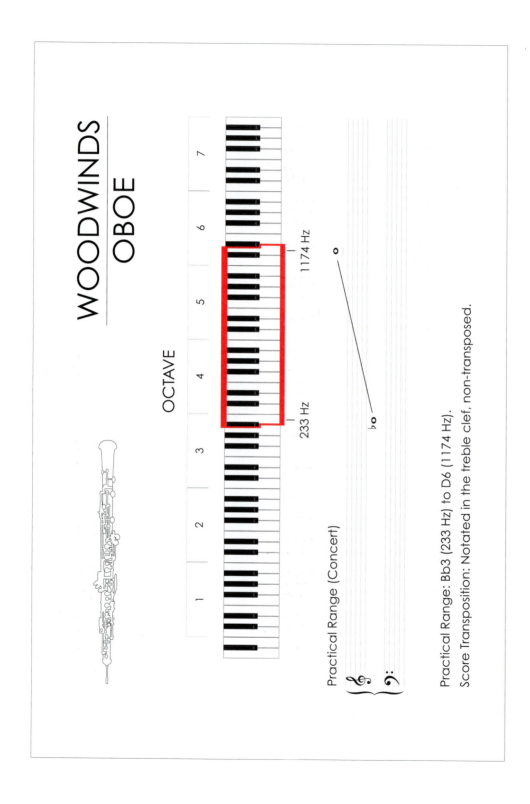

OCTAVE

233 Hz

1174 Hz

Practical Range (Concert)

Practical Range: Bb3 (233 Hz) to D6 (1174 Hz).

Score Transposition: Notated in the treble clef, non-transposed.

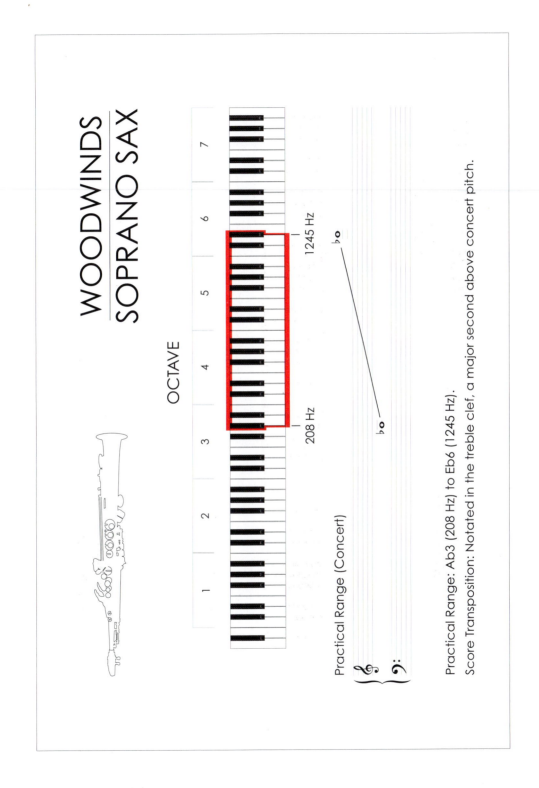

WOODWINDS
SOPRANO SAX

OCTAVE

Practical Range (Concert)

Practical Range: Ab3 (208 Hz) to Eb6 (1245 Hz).

Score Transposition: Notated in the treble clef, a major second above concert pitch.

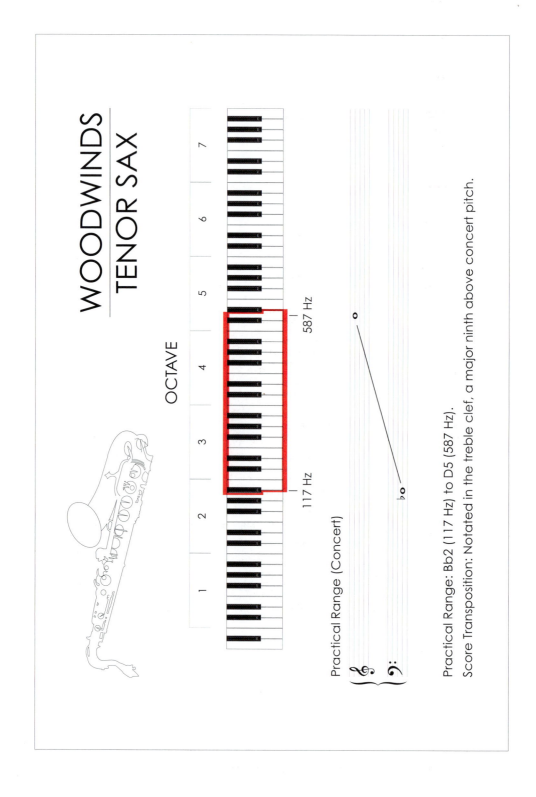

WOODWINDS
TENOR SAX

OCTAVE

1 2 3 4 5 6 7

117 Hz   587 Hz

Practical Range (Concert)

Practical Range: Bb2 (117 Hz) to D5 (587 Hz).

Score Transposition: Notated in the treble clef, a major ninth above concert pitch.

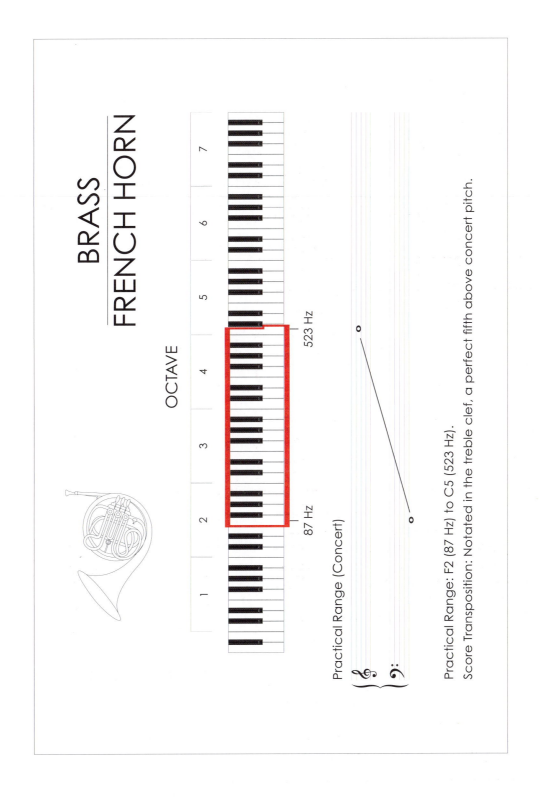

BRASS
FRENCH HORN

OCTAVE

523 Hz

87 Hz

Practical Range (Concert)

Practical Range: F2 (87 Hz) to C5 (523 Hz).

Score Transposition: Notated in the treble clef, a perfect fifth above concert pitch.

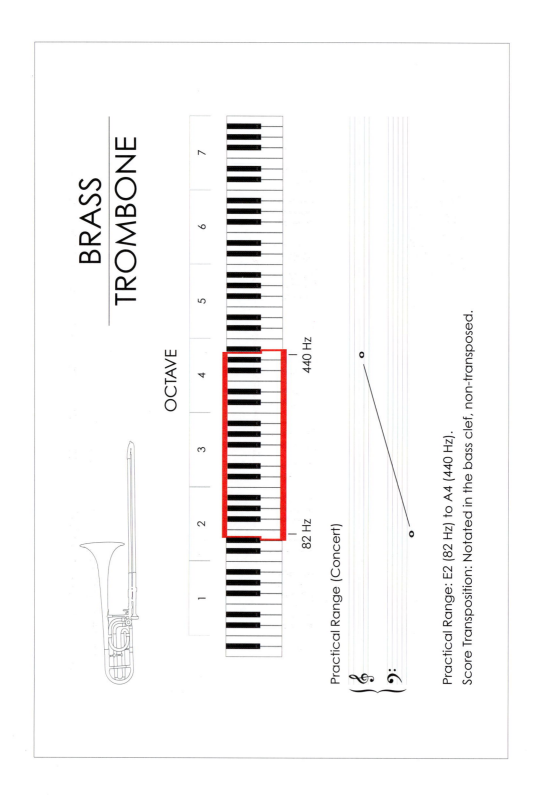

BRASS
TROMBONE

OCTAVE

1 2 3 4 5 6 7

82 Hz   440 Hz

Practical Range (Concert)

Practical Range: E2 (82 Hz) to A4 (440 Hz).

Score Transposition: Notated in the bass clef, non-transposed.

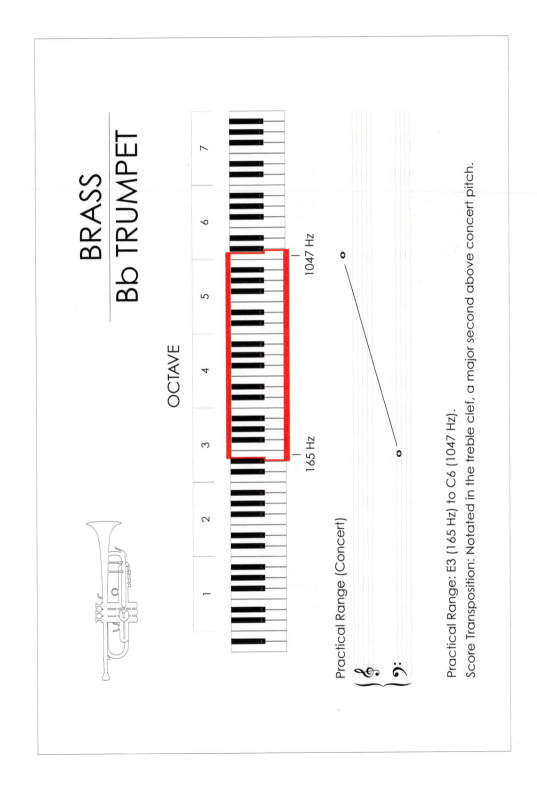

BRASS
Bb TRUMPET

OCTAVE

1   2   3   4   5   6   7

165 Hz

1047 Hz

Practical Range (Concert)

Practical Range: E3 (165 Hz) to C6 (1047 Hz).

Score Transposition: Notated in the treble clef, a major second above concert pitch.

BRASS
TUBA

OCTAVE

Practical Range (Concert)

Practical Range: F1 (44 Hz) to F4 (349 Hz).
Score Transposition: Notated in the bass clef, non-transposed.

PERCUSSION
GLOCKENSPIEL

OCTAVE

8272 Hz

784 Hz

Practical Range (Concert)

Practical Range: G5 (784Hz) to C8 (8272 Hz), or F5 to C8, F5 to F8, C5 to C8.
Score Transposition: Notated in the treble clef, two octaves below concert pitch.

159

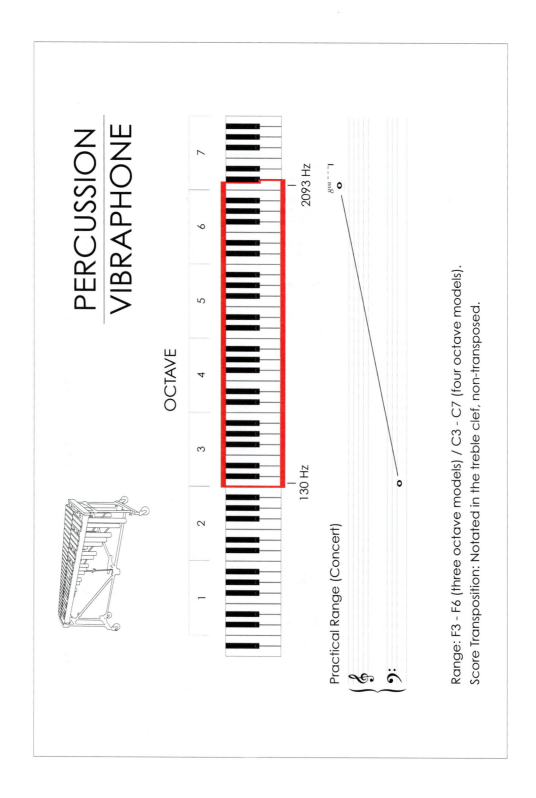

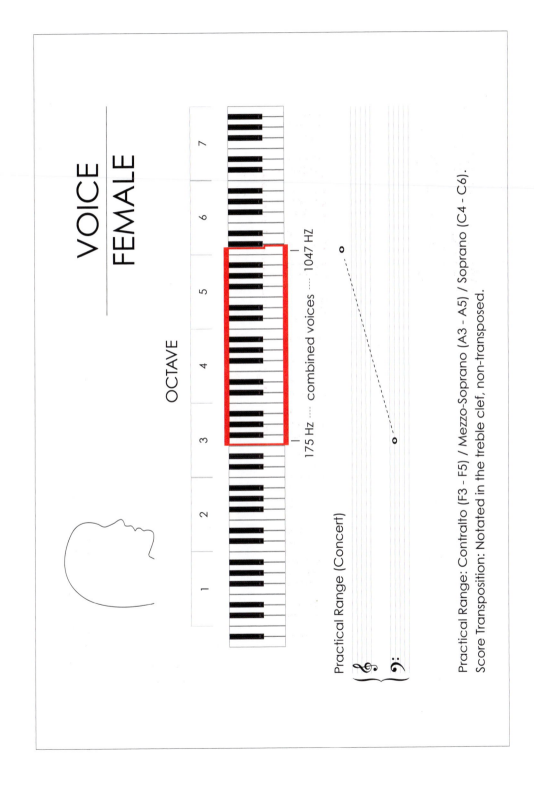

VOICE
FEMALE

OCTAVE

175 Hz ···· combined voices ···· 1047 Hz

Practical Range (Concert)

Practical Range: Contralto (F3 - F5) / Mezzo-Soprano (A3 - A5) / Soprano (C4 - C6).
Score Transposition: Notated in the treble clef, non-transposed.

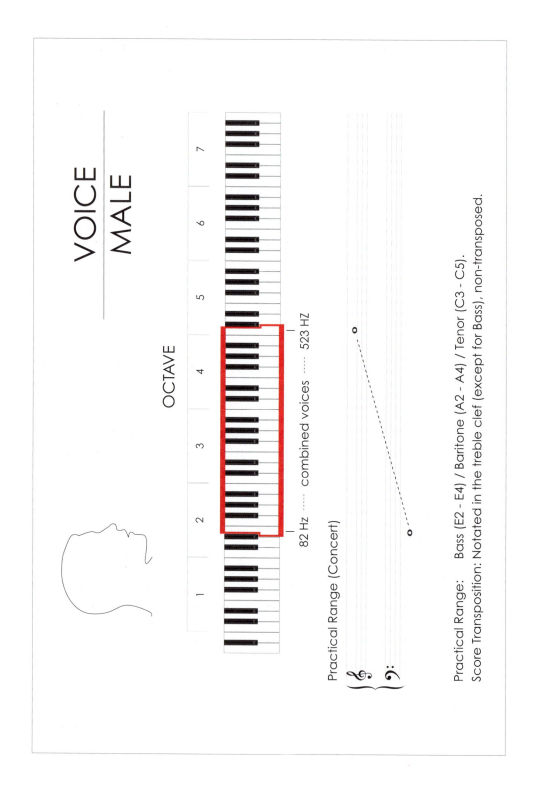

VOICE
MALE

OCTAVE

1   2   3   4   5   6   7

82 Hz ----- combined voices ----- 523 Hz

Practical Range (Concert)

Practical Range: Bass (E2 - E4) / Baritone (A2 - A4) / Tenor (C3 - C5).
Score Transposition: Notated in the treble clef (except for Bass), non-transposed.

THE 'RECCE' / VISITING THE RECORDING VENUE

It is vital for the production team to visit the recording venue for their impending sessions in advance. This action, commonly referred to as the 'recce', can help save a lot of time and avoid embarrassment, as in some cases a venue may be deemed unsuitable for sessions on personal inspection, e.g. contrary to advertising it may not accommodate the number of musicians required comfortably, etc.

From excessive extraneous noise to the lack of infrastructure, several reasons may make a location inappropriate for music production. On this basis, engineers/producers must consider all possible issues that might affect the result of sessions and discuss these with the rest of the team, e.g. artists, A&R, etc. Some of the aforementioned issues include:

- Unsuitability of equipment
- Difficulty in access (transport)
- Noise level constraints
- Size limitations
- Restricted access to mains power
- Environmental conditions, e.g. ambient noise, air conditioning, etc.
- Health and safety issues
- Lack of services, e.g. catering, parking, Internet access, etc.

It is important to note that in some cases what initially appears to be a hindrance may not have a negative effect on production, e.g. nature-related extraneous sounds that may become the 'sonic imprint' of a record.

Size Considerations

Recordists must examine session details regarding personnel closely, as these frequently dictate specific requirements regarding venue layout, size, etc. As an example, sessions involving classical musicians may surprise engineers that normally work with pop music, as the performers might require a greater amount of 'personal space' than what might be expected (a minimum of approximately 2 to 3 square metres per musician).

Other key aspects that are affected by venue size include the isolation between sound sources, reverberation time, etc. and all of these factors must be investigated and considered during the 'recce'.

EQUIPMENT REQUIREMENTS

The process of establishing the equipment requirements for a project should start with the final agreement over the arrangement of songs. At this stage, instrumentation related notes may be translated into technical audio signal information, outlining the tools required for production.

Signal Types

The following is an overview of the different signal types that recordists may work with during production:

Microphone Signals
- Origin: Voice; string, woodwind, brass and percussion instruments and backline amplification captured via microphones
- Nature of signal: Low-level balanced microphone signals
- Characteristics:
  - Very low voltages (microvolts to millivolts)
  - Low output impedance (commonly 600 ohms or less)
  - Microphone preamplification required

Electric Instrument Signals
- Origin: Electric guitars/basses, electric pianos, etc.
- Nature of signal: Unbalanced instrument level signals
- Characteristics:
  - Very low voltages (microvolts to millivolts)
  - High output impedance
  - Commonly recorded via direct injection (DI) or connected to backline amplification, which in turn is captured as an acoustic source, i.e. via microphones

Electronic Signals
- Origin: Synthesisers, sound modules, samplers, drum machines, sound cards, etc.
- Nature of signal: Balanced or unbalanced line level signals
- Characteristics:
  - Higher voltages (commonly exceeding 1 volt)
  - Low output impedance
  - Such signals may be fed directly into the line input of recorders

Electronic Sources and MIDI Signals

Some electronic instruments generate MIDI signals. The latter differ from audio signals, as MIDI is a data protocol that uses voltages to convey binary messages. MIDI makes it possible for musicians to record electronic parts quickly, i.e. without worrying about timbre, while allowing for a great degree of flexibility during editing and mixing.

Although electronic performances invariably involve the conversion of a MIDI stream into audible signals (as musicians need to hear what they are playing), it is not mandatory for engineers to record the audio that corresponds to a MIDI-related performance. Still, MIDI data should ideally be captured with accompanying sound, i.e. if viable, electronic performances should be recorded as audio as well as MIDI tracks. This will allow for the selection of 'patches' (sounds) and extensive editing at later stages, while providing a form of production 'insurance' in cases where original tracking sounds are considered best and are not replicable.

The Audio Tool List

When all the audio signal related information is determined (based on instrumentation), the data must be compiled and checked against the list of audio tools that will be available during sessions. Whether working in professional, semi-pro or home studios, recordists should always assemble an inventory of the devices that they will have access to during production, as such a register should help the planning process and increase efficiency. The aforementioned list must ideally include:

- Instruments / Backline amplification
- Microphones
- DI boxes
- Console(s)
- Microphone preamplifiers
- Equalisers
- Dynamic range processors, e.g. compressors and gates
- Effects processors, e.g. reverb and delay units
- Multitrack recorders / DAWs
- Two-track recorders
- Monitoring system, e.g. amplifiers, loudspeakers, etc.

> **Equipment Rental**
> It is common for large recording projects to require the rental of equipment, e.g. instruments, amplifiers, microphones, etc. These should be sourced from a trusted supplier and added to the audio inventory list.

THE SESSION PLAN

The session plan should be the final product of the planning stage, summarising all relevant information regarding production requirements and scheduling. This document must ideally include:

- Date, time and nature of sessions, e.g. tracking, overdub or bouncing
- Venue
- Suggested song(s) to work on (with key and tempo)
- Personnel and instrumentation breakdown (including call times)
- Engineering notes, e.g. click-track, backing track, general musician placement, music and microphone stands, number of cue mixes, etc.
- General requirements, e.g. media, catering, 'atmosphere'-related items, e.g. candles, rugs, etc.

It is important to note that under pressurised conditions, e.g. a busy production house or studio, the session plan may never be committed to paper, existing only in the producer's mind (or diary). It is nevertheless important to remember that planning equates to control and even in extreme conditions individuals can benefit greatly from mapping out their strategies.

Session Schedule / Date Planning

The scheduling of recording dates is one of the most challenging steps of session planning, requiring producers to be extremely organised and sensible when making projections. Ideally, only after extensive forethought should a schedule entry be made and backup plans are indispensable if costly studio time is at stake.

The allocation of funds or the breakdown of the budget between the different stages of production is dependant on a number of variables, e.g. how prepared the performers appear to be, etc., and figures will change as projects develop. Producers should always work with the most current, updated version of the budget and must be prepared to react immediately to situations as they arise. It is important to remember that projects must be finalised within their given budget and in many cases slight overestimation of costs may be advisable, i.e. the production team should err on the side of caution, without being unrealistic in their planning.

While it may not be practical to setout fixed, ideal time-related guidelines for the scheduling of sessions, it is certainly possible to split the production process into stages, allowing for resources and time to be allocated proportionally.

The following is an example of how the recording phase may be split:

1. Basic Tracking
   Loud rhythm instruments, e.g. drums, bass, electric guitar, electric keyboards, synthesisers, percussion (and possibly 'scratch vocals' for navigation and 'feel')
2. Overdubbing
   - Quieter rhythm instruments, e.g. piano, acoustic guitar
   - Lead vocals
   - Backing vocals
   - Instrumental solos
   - Special sound effects, hand percussion
   - Strings, horns, woodwinds, etc.

Upon careful examination of the steps necessary for the completion of a given project, it should be possible to estimate how the budget is to be distributed and one of the roles of the producer is to ensure that artists stick to the plan, while minimising the impact of finances on creativity.

Planning the Sound-Check

The inventory of audio tools allows for a sound-check signal path chart to be drafted. This preliminary plan of action should list suggested equipment chains for the different signals to be recorded, e.g. possible combinations of microphones and preamplifiers for acoustic sources. The selection of devices to be interconnected should be based on technical data and familiarity with the equipment, and it should provide the team with enough variety and/or backup options. The following is an example of a sound-check signal path chart for a singer-songwriter vocal and piano recording:

A – Main Vocals

Microphones	Preamplifiers
AKG C12	Neve 1073 DPA
Neumann U87	API 512C
Shure SM7	Focusrite Red 1

B – Grand Piano

<u>Microphones</u>	<u>Preamplifiers</u>
Neumann KM184 (x2)	Millennia HV-3C
AKG C414 (x2)	Manley Dual Mono
Earthworks QTC50 (x2)	Focusrite Red 8

From the preceding example it is possible to gather that the producer wants to try recording the main vocals and the piano using three different models of microphones and preamplifiers. These should be tested in all possible combinations, unless at some point of the audition stage a given chain is deemed ideal and further testing appears unnecessary.

A sound-check plan can help streamline the initial stages of the recording process, especially if an assistant is able to setup all microphones and preamps for A/B comparison prior to the beginning of sessions. This should allow the production team to spend more time auditioning signals and to focus on the aesthetic evaluation of sounds.

As producers combine all the knowledge gathered throughout the pre-production stage and finalise the session plan, the team should be ready to start production and enter the recording environment. At this stage, a few reminders can be extremely helpful and save all involved time and money. The following are some important final instructions for the performers:

- Be well prepared and warm up before sessions (do not rely excessively on inspiration).
- Practise using headphones and with a click-track (if sessions will require their use).
- Try to be focused and minimise possible distractions (do not invite unnecessary people to sessions).
- Change strings, drum heads, etc. ('break them in'), purchase spares and check instrument intonation, oil moving parts, etc.
- Check communication channels frequently, e.g. phones, email, etc.

Live Sound Technical Riders vs. Studio Session Plans
Recordists can benefit from analysing live concert technical riders, as such documents can help them create efficient and comprehensive session plans. Good technical riders commonly incorporate all relevant information regarding personnel, equipment, stage plans, etc., and as critical documents they must be assembled with extreme caution and attention to detail.

MUSIC PRODUCTION

RECORDING

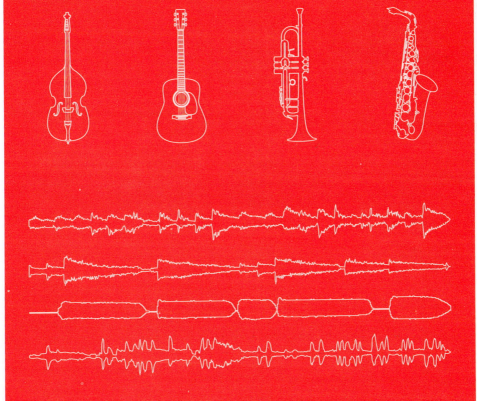

PRODUCTION

STUDIO PREPARATION

The first days of production are commonly very exciting, as hopefully a team will have worked hard during the weeks preceding their entrance into the recording environment. It is important to approach the set-up stage efficiently, but without the fear of changes in the plan. The quality of the technical team's work during tracking will potentially have the greatest impact over that of the final product, so standards should be set as high as possible. With this in mind, recordists should evaluate all the elements that may influence the result of sessions and attempt to optimise them.

ACOUSTICS

It is not uncommon for less experienced recordists to be only partially aware of the influence of acoustics over the quality of their productions. In certain cases this also seems to extend to professionals, as some engineers appear to overlook the possibility of varying the properties of the studio environment in order to achieve better results. It is arguable that the relationship between sound recording and acoustics has changed significantly since the introduction of DAWs and plug-ins, as these have made it easier for engineers to manipulate or even replace recorded material. Still, there is a limit to what may be achieved through artificial processing during editing and mixdown and few will question the value of working with strong raw materials, e.g. natural and authentic sounding recordings.

It seems advisable for those aiming to develop their recording skills to spend some time studying the fundamentals of acoustics, focusing their attention on the behaviour of sound in enclosed spaces. Although the aforementioned topic is undoubtedly complex, the knowledge of some of its basic principles may be sufficient to help recordists make simple and positive decisions in studio. As a starting point, engineers may choose to investigate the concepts of sound transmission, reflection and absorption, as this will help them to better understand what takes place in recording rooms.

When sound encounters a barrier, one or more of the following phenomena may occur:

Transmission
The energy of a sound wave can be transferred between environments, even when these are separated by a boundary. The mass and the density of the boundary influence this transfer and the magnitude of transmission is directly proportional to the amplitude (intensity) and inversely proportional to the frequency of the travelling sound wave. For these reasons, isolation is very difficult to obtain in the case of high

level, low-frequency rich sounds, e.g. bass drum, bass guitar, etc.

The transfer of energy through barriers is of particular relevance to those attempting to work in residential settings, where the presence of neighbours may make the recording and monitoring of loud sources impractical or even impossible. Contrary to what some inexperienced recordists think, the sole use of absorbing materials, e.g. foam, can do little regarding isolation and may simply alter the recording environment's frequency response negatively (through the dampening of high frequencies). A better alternative for those aiming to isolate their environment from the outside world is to increase the mass and the density of their boundaries (and their number), while aiming to decouple them from each other.

Isolation and Decoupling

The transmission of sound between environments can be minimised through decoupling. This requires a reduction in physical contact between the areas to be isolated and in many instances can only be achieved effectively through structural changes, e.g. the floating of floors, ceilings, etc. Although airborne noise is responsible for most transmission-related problems, the dampening of mechanical vibrations is very important when isolation is needed and recordists should be ready to work with materials such as neoprene rubber when aiming to stop mechanical transmission, e.g. creating platforms for amplifiers, etc.

Sound loses a lot of its energy as it travels through different media and this can be explored in circumstances where extended attenuation is needed, e.g. two barriers may be significantly more efficient in stopping sound if separated by an air gap as opposed to placed back to back.

Speaker Decoupling

Monitor pads can improve the quality of reproduced sound by reducing the interaction between speakers and the surfaces on which they rest. Such products are commonly made out of rubber and represent a small investment that may yield considerable results.

Reflections

When a sound wave meets a solid obstacle of significant size it will have some of its energy deflected and sent back in the direction of origin. Once again the composition and dimensions of the barrier will dictate how much of the energy will be redirected and how much will be transferred. This phenomenon has great importance in the case of musical content, due to resulting alterations of amplitude and timbre as direct and reflected sounds combine.

When numerous 'versions' of a signal (echoes) are generated in succession due to the presence of boundaries, sound waves will follow multiple paths and eventually interfere with one another. If the pre-delay gap (the time lapse between direct sound and first reflections) is small, the aforementioned interference will affect the timbre of signals noticeably (due to comb filtering).

Good sounding natural reverberation is the sum of complex reflections that occur under favourable conditions. On the other hand, non-flattering reverb, e.g. flutter echo, is almost always available, although such effect is commonly detrimental to musical recordings and, if possible, best avoided.

Negative Interference / Comb Filtering

The influence of early reflections over the timbre of sound can be demonstrated through real-time experiments. As an example, if a recordist holds a small, battery-operated speaker producing white noise and approaches a wall, a 'phasing' like effect will be detected (a manifestation of comb filtering). Such phenomenon illustrates why engineers should place performers away from boundaries and why they should consider the placement of reflective surfaces in the recording environment carefully, e.g. music stands, etc.

Standing Waves

The presence of parallel walls in small, enclosed spaces accounts for the pronounced occurrence of standing waves. These result from reoccurring, predictable reflections and translate into a frequency spectrum that changes in relation to position, i.e. small rooms with

Standing Waves (continued)

parallel surfaces present a very inconsistent spatial response. Unfortunately this is the reason why a lot of recordists are not able to achieve satisfying results in their home studios, as most untreated domestic rooms present peaks (resonances) and valleys in frequency response and do not offer a trustworthy critical listening position.

Standing Waves in Control Rooms

Control rooms should be free of pronounced standing waves. If the latter are not avoidable due to construction-related constraints, reflections must be 'tamed' and reduced to a minimum through absorption and/or diffusion. There are a few ways to assess the impact of standing waves over the sound quality of a given control room. One of the simplest procedures involves the playing and auditioning of ascending pure tones at constant levels, e.g. engineers may set up an oscillator or an instrument to play 41 Hz (E1), 44 Hz (F1), 46 Hz (F#1), etc., checking for audible inconsistencies at the critical listening position.

NB It is important to take the concept of 'equal loudness' into consideration during this experiment (see Appendix 3).

Resonances (Standing Waves) in Live Rooms

Although commonly described as harmful, resonances may be explored artistically in live rooms if the 'colouring' of an original source is desired, e.g. a 'bassier' production of sound from a 'thin' emitter. As an example, in home recording environments, a sound source may be placed in a highly resonant space such as a bathroom or a room's corner for effect (although this should be approached with caution).

Diffusion

Diffusion is the name given to the 'scattering' of sound through reflections that are theoretically random in orientation, i.e. ones that do not follow a predictable travel path. This phenomenon may help to remedy the problem of comb filtering and flutter echoes, making small rooms sound more 'musical' or less 'boxy'.

Diffusers

Diffusers come in different shapes and sizes. Some look very complex, e.g. skyline models, while others seem much simpler, e.g. QRDs (quadratic residue diffusers). The theory supporting the design of such devices is based on advanced mathematics and although they may seem difficult to build, recordists with a little experience in carpentry should be able to assemble a basic diffuser, which could help improve the sound quality of their listening environment.

Absorption

The energy of a sound wave may be 'absorbed' or transformed through friction or 'dampening'. This justifies the use of porous materials, e.g. acoustic foam, to control the transmission of sound and numerous acoustic devices using this principle have been developed to help 'tame' reflections in problematic environments.

The conversion of acoustic energy into heat is most efficient when a sound wave travelling at maximum speed meets a porous obstacle. To achieve such a degree of absorption, engineers may:

- Employ a free-standing porous boundary with a thickness that matches half the wavelength of the lowest frequency to be absorbed, e.g. a 1.56 metre-thick absorber in the case of 110 Hz (A2).
- Use a thinner absorber placed at one quarter of the same wavelength away from a reflective surface, e.g. a less thick porous boundary placed at 0.78 metres from a wall in the case of 110 Hz (A2).

Both aforementioned strategies guarantee that a lowest 'target' sound wave and subsequently other higher-frequency waves will have an instance of maximum velocity within the absorber.

Bass Traps
Low frequencies have long wavelengths, which challenge the use of porous materials for absorption, e.g. a free-standing boundary would need to be at least 4.2 metres thick in order to absorb 41 Hz (E1) efficiently (where 4.2 metres is half the wavelength of 41 Hz).

Bass traps are devices designed to capture low-frequency waves at points of build-up, e.g. corners. As we have seen, the placing of porous materials at a distance from boundaries extends their bandwidth of absorption, allowing for smaller devices to be used.

Strategically placed bass traps can be very efficient in the control of low-frequency energy.

Applied Variable Acoustics
The use of variable acoustics or portable acoustic devices allows for dynamic control over the sonic characteristics of an environment.

The following pages contain simple examples of variable acoustic concepts applied in practice.

VARIABLE ACOUSTICS

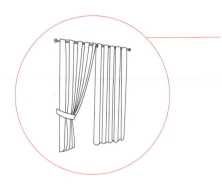

Absorptive Materials

One of the simplest ways to vary the acoustic properties of an environment is to increase or decrease the quantity of absorptive elements within it. As a simple example, heavy curtains may be installed in a recording space to be either drawn or pulled back depending on the amount of absorption desired. The same applies to rugs and pillows, which may also be used to control (high-mid and high-frequency) reflections.

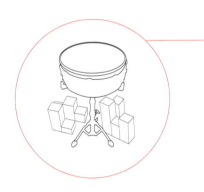

Portable Diffusers

The placement of irregular shaped objects or a small diffuser under a snare drum may make the instrument sound less 'phasey' or 'metallic'.

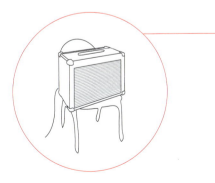

Amplifier Platforms

The raising of a guitar amplifier may help clear the low-mid range of the instrument considerably, as the sound source is decoupled from the floor. A small, open-back chair or table can be used efficiently for such purpose.

VARIABLE ACOUSTICS

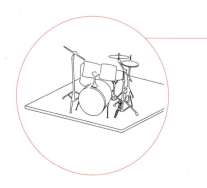

Reflective Surfaces

The placement of a reflective panel in the recording environment can alter the timbre of instruments making them sound more 'urgent' or energetic. As an example, a wooden panel placed under a drum kit may bring the instrument to life and lead a performer to play more lively.

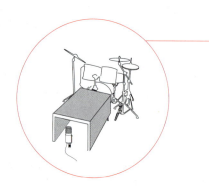

Isolation Tunnels

A bass drum tunnel can help recordists achieve a certain degree of isolation between key elements of a drum kit and gain control over the inner balance of the instrument, i.e. by removing or reducing the 'leakage' from the hi-hat and the snare drum from the bass drum channel, engineers may be able to rebalance the kit's levels during mixdown with ease. Bass drum tunnels can take the simple form of two stools or chairs placed on either side of the front of the bass drum with a blanket or duvet placed over them.

POWER AND DISTRIBUTION

The quality of power supply is a concept that is usually ignored by engineers working in well-established commercial studios, as the latter commonly invest considerable resources in the treatment of the power they utilise. The picture can be quite different in the case of recordings that take place in residential environments, where recordists may need to check and plan their access to mains supply, in order to avoid or minimise problems such as 'hum'.

It is not the objective of this text to cover the basics of electronics theory, as many publications dedicated to the topic are readily available. What follows are simple, basic suggestions for recordists working in amateur or semi-professional studios:

1. Measure mains power at all supply points (outlets) and ensure that values are consistent and do not fall beyond plus or minus 10 percent outside the stated requirements of the equipment being used, e.g. if a microphone preamplifier is rated at 240 Volts the mains supply should not approach the limits of 216 Volts and 264 Volts.

    NB NEVER attempt to perform step 1 if you have not been trained to use a voltmeter or if you do not own a standalone safe power outlet tester.

2. Ensure all audio-related equipment shares the same path to the ground (not necessarily the same outlet) and that such path is not shared by non-audio-related devices, such as light fixtures, portable heaters, kitchen utensils, etc.

    NB Simply reduce the use of non-vital, peripheral electrical devices if the electrical wiring of the recording environment is not clear.

3. Utilise surge-protection supply devices, e.g. specialised high-quality power strips, to feed all sensitive audio equipment, e.g. computers, audio interfaces, etc. (this may help avoid 'clicks' or 'spikes' in recordings).

The following page contains a graphic description of the process of measuring voltages.

MEASURING VOLTAGES

Select a Signal to Measure

In this example the output of a portable signal generator is being measured. The device is set to output a sine wave signal at + 4 dBu.

Make the O/P Connection

Check if the device provides balanced and/or unbalanced connections. In the example an unbalanced output on RCA connection is being used.

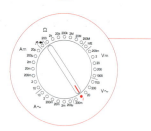

Set the Multimeter

Set the multimeter to measure AC voltages (commonly V~) and choose the appropriate range. The latter should correspond to the lowest value offered that is higher than the expected measured voltage, e.g. if working in the +4 dBu range set the multimeter at 2 Volts AC (the expected voltage is in the 1 Volt range).

Measure the Signal

Turn on the device to be tested and touch the centre and the screen of the connector with the red and black probes of the multimeter respectively. For this example, the multimeter should display a reading of 1.228 Volts AC (approximately).

EQUIPMENT ALIGNMENT

Following power-up, the different elements of the recording chain must be aligned to ensure they will interface with each other in a coherent, optimised fashion.

Analogue Alignment

The starting point of strictly analogue alignment is the establishing of standard operating levels and clipping points for all the elements in the audio path. In possession of this information, recordists can subsequently determine the headroom of all devices and set the 'standards' for that signal chain, i.e. the lowest clipping point should be taken as a path's 'ceiling' (unless distortion is desired). As an example, if a chain incorporates a console with a clipping point of + 18 dBu and a tape-based recorder that 'clips' at + 12 dBu, the latter value should be used as the ceiling of the system, i.e. operators should avoid boosting signals beyond + 12 dBu.

Digital Alignment

Digital alignment is the simplest of all three processes. Here, operators must only ensure that a common 'average' or reference level is selected and utilised consistently, which in some cases may require a simple change in DAW settings (preferences).

NB As far as the choice of the aforementioned reference level is concerned, two main standards are commonly used:
- EBU R68 = − 18 dBFS
- SMPTE RP155 = − 20 dBFS.

Hybrid (Analogue / Digital) Alignment

The alignment process is somewhat different when digital devices are used alongside analogue equipment. In digital audio technology, the fixed reference 0 dBFS does not represent an average, but a maximum value that must not to be exceeded (unless gross digital distortion is desired). This 'peak' value must therefore be equated to the maximum dBu or dBV value that can be handled by the analogue devices in the chain, without the production of objectionable distortion.

In hybrid alignment, the figures that correspond to the 'average' operating level and subsequently the headroom may in most circumstances be set by the user, i.e. professional converters allow for operators to set the dBFS level that will correspond to a counterpart analogue standard operating level, e.g. − 18 dBFS may be set to match + 4 dBu (1.228 Volts RMS) or 0 dBu (0.775 Volts RMS) on conversion.

CHANNEL LINE ALIGNMENT

Select a Tone Generator

Select a device capable of generating a stable 1kHz sine wave tone at +4 dBu (if operating at 'pro' level).
NB Some consoles incorporate a built-in oscillator.

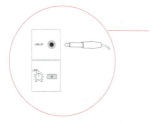

Connect the Device

Patch the output of the generator into the line input of a channel path.

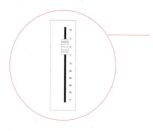

Adjust the Channel Fader

Set the channel path fader at unity gain (zero or 'U').
NB A pre-fader (PFL) solo may be used alternatively.

Meter the Signal

Set the meters to display channel levels and check that a reading of 0 VU or + 4 dBu is produced.

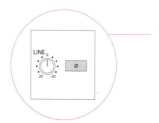

Adjust Gain if Necessary

Compensate for any possible discrepancies using the line gain control of the channel.

'HYBRID' ALIGNMENT (1)

Route the Signal to the DAW

After following the procedure described in the 'Channel Line Alignment' diagram, route the channel output to the DAW input (here the 'direct' output of Channel 1 is feeding the first input of an audio interface).

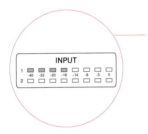

Meter the Converter Input

Use the audio interface meters to check the signal level at the input of the converter. Ensure that the chosen standard is met, e.g. + 4 dBu = – 18 dBFS.

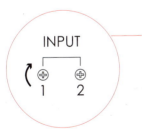

Adjust the Converter I/P Gain

Adjust the input gain of the converter if necessary (if such function is available).

Set the DAW to Record

Create a track on the DAW, setting its input and output appropriately. Set the track into record ready status, i.e. arm the track to record. NB Ensure the track's fader is set to unity gain.

'HYBRID' ALIGNMENT (2)

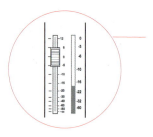

Meter the DAW I/P Signal

Meter the signal using the DAW track meters and confirm that the level is still consistent, e.g. – 18 dBFS.

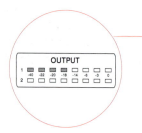

Meter the Converter Output

Use the audio interface meters to check the signal level at the output of the converter. Ensure that the chosen standard is maintained, e.g. – 18 dBFS.

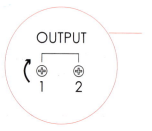

Adjust the Converter

Adjust the output gain of the converter if necessary (if such function is available).

Meter the Return Signal

If the output of the DAW is feeding a console, set the meters to display 'return' or 'from recorder' levels. Check that a consistent reading is still being produced, e.g. 0 VU or + 4 dBu.

Decibels

A decibel value is an expression of a ratio, or a comparison, in logarithmic form. Logarithms are used in engineering as they allow for a simpler, more manageable representation of very vast ranges. As an example, the human ear is capable of 'safely' handling pressure spanning approximately between 0.00002 and 63.25 pascals or between 0 and 130 dBSPL. This substantial (1.265 x 10^6) range seems more conveniently represented in dBSPL form, since the latter allows for faster communication and the easier graphic display of quantities, e.g. through cartesian plane graphs.

Standard Operating Levels

The concept of standard operating level (SOL) is used to describe a recommended average level at which equipment should run in order to operate well above the noise floor and without producing objectionable distortion. The following are industry standard operating levels:

Professional Equipment
0 VU = + 4 dBu = 1.228 VRMS

Semi-Professional Equipment
0 VU = − 10 dBV = 0.316 VRMS

NB Two other standard operating levels, 'broadcasting' (commonly + 8 dBU) and 'hi-fi' (− 10 dBu) are also used in the audio industry, although these may vary geographically.

PREPARING THE DAW SESSION FILE

Following the alignment of the recording equipment, technicians should prepare the DAW project file. The session plan should be consulted at this stage as it should specify:

- If a click track is necessary (and its tempo)
- If a backing track or sequence will be used
- Whether samples will be incorporated into the production.

Alternatives to Click Tracks

Some drummers may find the sound of a traditional click track uninspiring or annoying. In such cases, session engineers may:

- Add a beat-related, e.g. crotchet, delay to the sound of the drums in the cue mix only (which will help drummers detect when they are drifting out of time).
- Use a looped sample of a steady percussive 'groove', e.g. congas playing a few bars, as a guide to help the drummer keep a steady pulse.
- Use a track of cowbell, woodblock, tambourine, etc. played 'live'.

Time-Stretching and the Click-Track

Whenever possible, operators should record the audio signal used for time keeping with their relevant sessions, as such files may be invaluable in the case of a DAW or platform change (which could require the remapping of tempo, etc.). When using advanced time stretching, e.g. Elastic Audio, Flex Time, etc., recordists can easily alter the tempo of audio files used for time reference, by making these 'tick-based' and not 'sample-based'.

Sample and Tick-Based Timescales

Current DAWs allow for the position (or 'address') and length of regions to be given in samples or in bars, beats and tick values. When a sample-based timescale (timebase) is used, audio regions are positioned according to sample value, e.g. in 44100 Hz, a region placed at 'address' 44101 will play one second after the start of the session and this position will remain fixed regardless of tempo changes.

A tick-based timescale allows for the placement and speed of playback of MIDI and 'elastic' or 'flex' time audio regions to be automatically updated according to tempo, e.g. a region placed on the second quaver of the third beat of bar four and lasting for two bars will play at its given music-related position and for its full length at any selected tempo (speeding up or slowing down accordingly).

Sample and Tick-Based Timescales (continued)

In a tick-based scale, region positions are described in bars, beats and ticks, which correspond directly to the following musical note values:

- Semibreve (whole note) = 3840 ticks
- Minim (1/2 note) = 1920 ticks
- Crotchet (1/4 note) = 960 ticks
- Quaver (1/8 note) = 480 ticks
- Semiquaver (1/16 note) = 240 ticks
- Demisemiquaver (1/32 note) = 120 ticks.

Using a previous example, in common time or 4/4 and with the DAW 'division' set at 1/4 note, a region placed on the second quaver of the third beat of bar four will have the address 4 | 3 | 480.

Session File Templates

Recordists can benefit from the use of templates by not having to create multiple tracks, assign track inputs and outputs, create busses (cues, effects), assign buss inputs and outputs, etc. for each song to be recorded. DAWs allow users to save session file templates, which can be very useful when numerous songs with similar set-ups will be recorded in succession.

Auto-Backup

Recordists should enable 'auto-backup' or any other automatic background saving function of their DAW as soon as possible and ideally before sessions. This may allow some work to be salvaged in the case of an application or computer crash.

Labelling Folders and Files

It is important for DAW users to be sensible and remain consistent when labelling folders and files. As a suggested approach, folders

Labelling Folders and Files (continued)

and sub-folders may be organised in the following hierarchical manner:

1. Main client, e.g. a record label or producer (top folder)
2. Artist (sub-folder of 'Main client')
3. Album or project (sub-folder of 'Artist')
4. Song (sub-folder of 'Album'), e.g. Song X
5. Session Files (sub-folders of 'Song'), e.g. Song X_01JAN13_A or Song X_010113_A (following the format Song_Date_Version).

Slating the Recorder

It is important to document the operating level used for each recording session, as this may vary between productions. The process of 'slating' the recorder, i.e. the recording of tones at reference level, is arguably seen as less important since the introduction of digital recording, although this procedure still plays a vital role in the alignment of devices during the subsequent stages of production.

In cases where slating is not possible, e.g. if an oscillator or signal generator is not available, DAW operators should at least note down the selected standard for each session, e.g. + 4 dBu = − 18 dBFS.

Audio File Types

It is recommended that recordists work with or generate (data) uncompressed audio files with wide support, i.e. files that are known for their resilience and mobility. Currently, the Broadcast Wave (.bwf) format is considered to be the most universal, being supported by a variety of platforms and applications. Other acceptable file types include: .aiff and .wav, while compressed or not currently supported file types, such as .mp3, .m4a, .m3u, .wma, .ra, .sd, .sd2, etc., should be avoided during production.

LIVE ROOM SET-UP

Following, or simultaneously to, the powering-up and calibration of the recording equipment and, if required, the creation of DAW session files, recordists should prepare the studio for the arrival of the performers. At this stage, whoever is responsible for the set-up (possibly the assistant engineer) should consult the session plan in order to gather information regarding:

- Musicians' placement, including the use of microphone stands and baffles
- Transducer and backline selection and positioning (including rental-related information)
- Cabling
- Inputs and track allocation
- Headphone / Foldback mix and talkback requirements
- The distribution and arrangement of comfort or atmosphere-related items
- Instrument-related requisites, e.g. oiling of pedals, tuning of drums, etc.

The set-up stage provides an opportunity for recordists to start their relationship with artists and producers on positive terms and it is important to note that strong first impressions commonly have a long-lasting impact in music production environments.

INSTRUMENT PLACEMENT

Spatial positioning can affect the sound of instruments dramatically so recordists should be very careful when selecting the placement for their sound sources. In rock and pop music, it is common for engineers to start the set-up process by searching for the ideal position for the drum kit, which seems sensible as this instrument is usually the largest and most complex element in standard productions.

Some of the most common techniques used for drum placement involve the playing and auditioning of a given component of the kit at different positions within the recording environment. Some individuals use the floor tom or the bass drum as the 'placement testing' instrument due to their low-frequency resonance. With that in mind, if a pronounced low-end resonance is desired, in most small to medium-sized, rectangular-shaped rooms, the best sounding position will almost certainly lie in the centre of all dimensions, i.e. the middle of the room.

Considering the importance of the snare drum in contemporary rock music, recordists may alternatively use this instrument for their drum kit placement tests. In such case, a technician may easily carry and play a snare drum around the studio unaided, listening for the interaction between the instrument's 'dry' sound and its early reflections. Once the most appropriate sound (balance) for the 'backbeat' is found, the drum kit may be assembled around it.

There are no hard rules on how to place the rest of the performers and respective backline in relation to the drum kit, although it is common for bass players to prefer the hi-hat side of the kit and to request a line of sight with the drummer's bass drum foot. Assistants should always consult the session plan for instructions before making any decisions involving placement.

Upon the arrival of the performers at the recording space, technicians should be ready to start the sound-check process. At this point, the backline should be in its general position (session plan) and drum kit should be in its place and in tune (see Appendix 6 for a description of drum tuning procedures).

The sound-check can be long and arduous process that demands concentration and it should be free from any unnecessary distractions relating to set-up.

Artist Comfort

Whenever possible and reasonable, the well being of the artists should be given priority over all else. This includes the allocation of adequate space for each of the performers and the strategic placement of baffles, allowing good visibility to be maintained throughout (encouraging interaction / interplay). The importance of artist comfort should never be underestimated, as musicians may perform beyond expectations when inspired by their surroundings.

Comfort and atmosphere-related items may include:
- Tapestries and pillows
- Chairs and small tables
- 'Mood' lighting devices and candles
- Artwork
- Incense and other aromatherapy products.

MICROPHONE PLACEMENT

Microphone placement strategies can range from the scientific to the borderline superstitious. Some technicians seem to enter the live room with a clear idea of where to place all transducers and are ready to accept immediate results as final. Others appear to thrive in the quest for the best possible microphone position and will not rest until they are completely satisfied with the results. Independent of the approach, the recording team should aim to capture sounds that will fit each production distinctively, even if this requires the recording of sounds that do not appear perfect in isolation.

Although the knowledge of instrument radiation pattern and familiarity with microphones and their qualities can undoubtedly help recordists make important decisions and save time in the studio, it is unreasonable to expect a certain formula to work consistently for the recording of a given sound source. With this in mind, microphone placement should be viewed as a dynamic process that demands highly developed listening skills and recordists should not be afraid to spend a considerable amount of time in the live room examining the natural sound of instruments at different angles and distances.

Evaluating Sound

Engineers use different techniques for the evaluation of the sounds they will record. Some start by relying exclusively on their ears while others prefer to take the transducer (microphone) into consideration from the beginning of the process. Both approaches have advantages and may be better suited for certain purposes, e.g. short or long sessions, quiet or loud instruments, etc.

1. The 'Naked Ear' Approach
In this approach, while a performer plays an instrument, recordists obstruct or cover one of their ears (if recording in mono) and move their heads within the source's surrounding area. Upon finding the spot where the instrument sounds best, the most appropriate microphone is selected to preserve or enhance (complement) the sound at the given position.

It is advisable for technicians to start this procedure by moving their heads within two dimensions, i.e. left / right and up / down, at a predetermined distance from the sound source. As the best position in relation to height and side displacement is established, recordists may move closer and further away from the source, evaluating the resulting change in spectrum ('blend' of frequencies) and the suitability of added reverberation.

It is imperative for technicians to be *extremely* careful when auditioning loud sound sources, e.g. drums, guitar cabinets, etc., as some instruments can produce sound pressure at harmful levels. In the case of 'spot' microphones, the 'naked ear' approach *must never be attempted* in close proximity to any component of a drum kit.

2. The 'Microphone-Based' Approach

Some individuals prefer a microphone-based approach to sound evaluation, as they choose to audition instruments as they are picked up by the transducer. This technique requires the use of isolating headphones (closed back or 'ear defender' style) and may demand technicians to listen to signals at a considerably loud level, in order for the headphone output to be louder than the natural sound in the room.

It is not advisable for technicians to instruct performers to play quietly in order to facilitate the evaluation process, as no instrument will sound harmonically identical when played at quiet and loud levels, i.e. artists should play the part that will be recorded at its appropriate intensity (dictated by the composition).

In this approach, the microphones are commonly predetermined, e.g. by a session plan, and are moved within the sound radiation area (in the same way as the recordist's head is moved in the 'Naked Ear' approach).

The following page contains a depiction of both the 'naked ear' and the 'microphone-based' approaches to microphone placement.

THE 'NAKED EAR' APPROACH

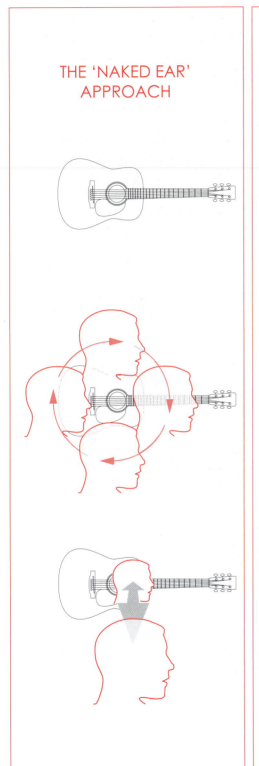

THE MICROPHONE-BASED APPROACH

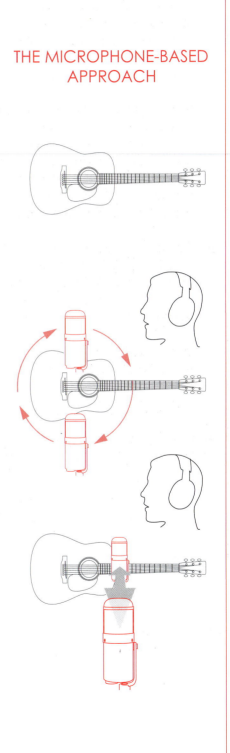

Microphone Set-up Procedure

1. In the control room:
   - Make sure that phantom power is off in all console channel strips and/or standalone preamplifers.
   - Bring all console faders down or mute all channels.
   - Mute the control room loudspeaker outputs.

2. In the live room:
   - Gather the necessary number of stands, plus a few spares.
   - Check every stand part, e.g. clamps, boom arms, etc., and make sure that all joints can be tightened securely.
   - Ensure that only rubber-covered parts come in contact with the floor (if necessary place foam pads under stands).
   - Place stands in the general area and at the approximate height at which they will be used.
   - Run cables from the input wall box (unpatched) to the microphone stands avoiding close proximity with power cables (if signal carrying cables must cross paths with power leads, they must be set and 'gaffa' taped at a 90° angle – minimising interference).
   - Attach all clips that correspond to the microphones that will be used to their respective stands.
   - Bring one microphone at a time to their corresponding stand (keep fragile microphones inside their protective bags).
   - Place each microphone securely onto their clip / stand.
   - Check that stands will not 'droop' with the weight of microphones.
   - Patch the female XLR connector firmly onto the microphone and the male XLR connector onto the wall input box.

3. In the control room:
   - Turn phantom power on where required.

The following pages contain instrument-related information that may help recordists with microphone choice and placement procedures.

STRINGS
BASS

NOTES

The double bass:

- May produce high sound pressure levels if played with a bow (in excess of 100 dBSPL at close proximity).
- May sound overtly aggressive at close proximity (when bowed).

SOUND RADIATION

No true omnidirectional range.

Slight low-mid range bias

Directive towards the front

COMMON RECORDING TECHNIQUES

Pizzicato (fingers)

Commonly recorded off-centre, i.e. to the left or the right of the bridge.

STRINGS
CELLO

NOTES

Violloncelos or cellos:

- May produce high sound pressure levels (approximately 100 dBSPL at 1 metre).
- May sound overtly aggressive at close proximity (commonly recorded at some distance).

COMMON RECORDING TECHNIQUES

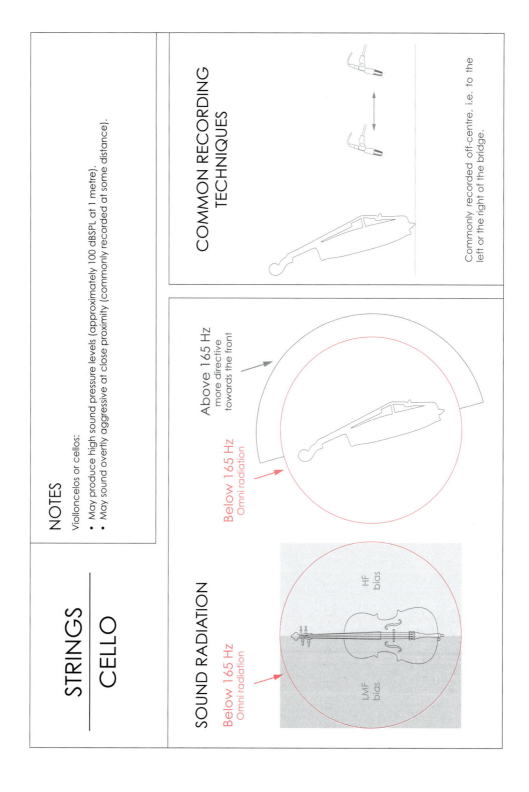

Commonly recorded off-centre, i.e. to the left or the right of the bridge.

SOUND RADIATION

Below 165 Hz
Omni radiation

Above 165 Hz
more directive
towards the front

Below 165 Hz
Omni radiation

HF bias

LMF bias

STRINGS
GUITAR

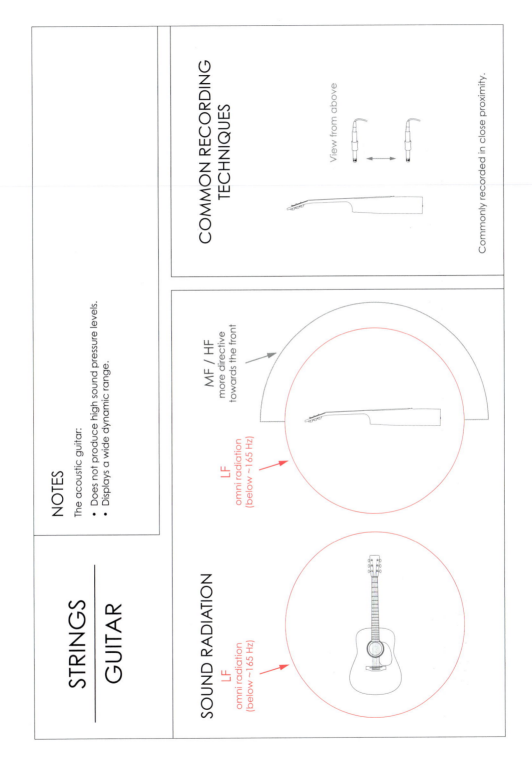

NOTES

The acoustic guitar:
- Does not produce high sound pressure levels.
- Displays a wide dynamic range.

COMMON RECORDING TECHNIQUES

View from above

Commonly recorded in close proximity.

SOUND RADIATION

LF
omni radiation
(below ~165 Hz)

LF
omni radiation
(below ~165 Hz)

MF / HF
more directive
towards the front

STRINGS

ELECTRIC GUITAR / BASS

GENERAL NOTES

Electric guitar / bass setups:

- Span from large stacks, combining amp heads and cabinets to small portable standalone amplifiers.
- May produce extremely high sound pressure levels.

COMMON RECORDING TECHNIQUES

Commonly recorded with one or more of the following:

- A dynamic, moving-coil microphone placed at very close proximity to a speaker driver.
- A ribbon microphone placed close to a driver (careful not to exceed the maximum SPL of the transducer – consider setting the microphone at an off-axis angle).
- A condenser microphone positioned at a distance, capturing a fuller sound that incorporates room reflections.

RECORDING NOTES

- The sound produced by a speaker driver varies considerably between the centre and the outside of the cone. Microphones aimed at the dust cap (centre) will pick up a much brighter sound than those aimed at the outer edge of the cone.

- The angle between the microphone and the driver also influences timbre. Recordists should experiment with aim, distances and angles.

- Some recordists may choose to capture the rear sound of open-back amplifier cabinets. Attention to polarity is necessary if the outputs of rear and front-facing microphones are summed, i.e. the polarity of one of the signals may need inversion.

- Multiple microphone signals may be combined and balanced during mixing, allowing for a variety of timbres to be achieved without the need for artificial equalisation.

- The acoustic sound of an electric guitar may be recorded and combined with the amplified signal. This can add definition to heavily distorted guitar sounds.

- The physical lifting of an amplifier, i.e. a decoupling from the floor, can clear the low end of guitar signals.

STRINGS
VIOLA

NOTES

The viola:

- May produce high sound pressure levels (in excess of 100 dBSPL at close proximity).
- May sound overtly aggressive at close proximity.

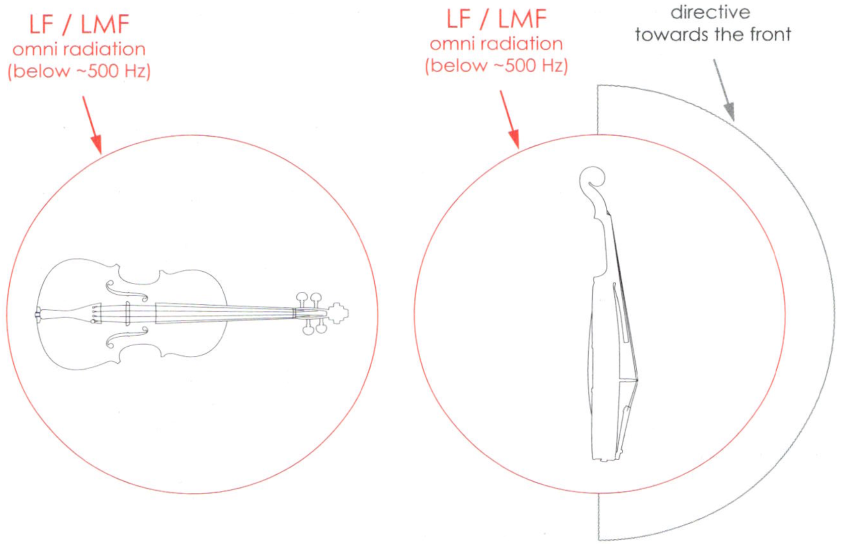

COMMON RECORDING TECHNIQUES

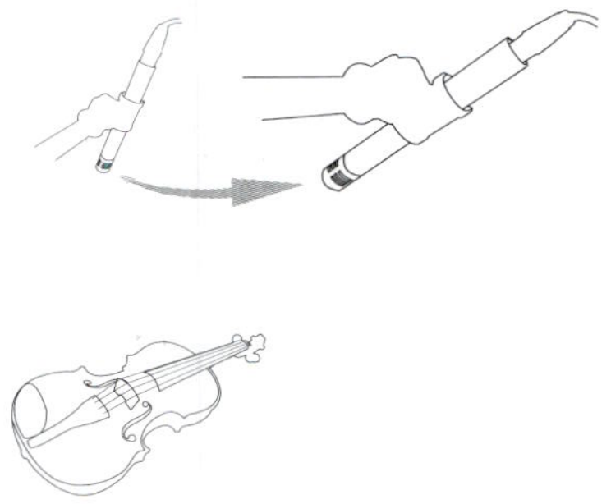

Commonly recorded from above or on-axis with the bridge, at a distance.

SOUND RADIATION

LF / LMF
omni radiation
(below ~500 Hz)

HMF / HF
directive
towards the front

LF / LMF
omni radiation
(below ~500 Hz)

STRINGS
VIOLIN

NOTES

The violin:

• May produce high sound pressure levels (in excess of 100 dBSPL at close proximity).
• May sound overtly aggressive at close proximity.

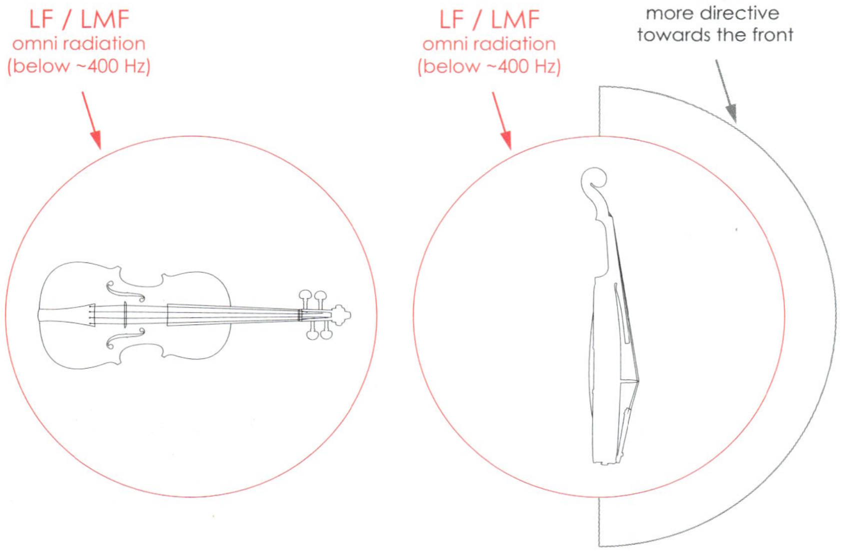

COMMON RECORDING TECHNIQUES

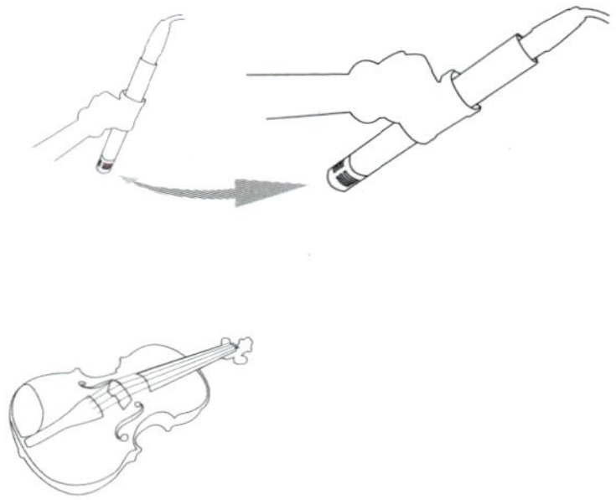

Commonly recorded from above or on-axis with the bridge, at a distance.

SOUND RADIATION

LF / LMF
omni radiation
(below ~400 Hz)

HMF / HF
more directive
towards the front

LF / LMF
omni radiation
(below ~400 Hz)

WOODWINDS
ALTO SAX

NOTES

The alto saxophone:

- May produce high sound pressure levels (in excess of 100 dBSPL in close proximity).
- May sound overtly aggressive on-axis.

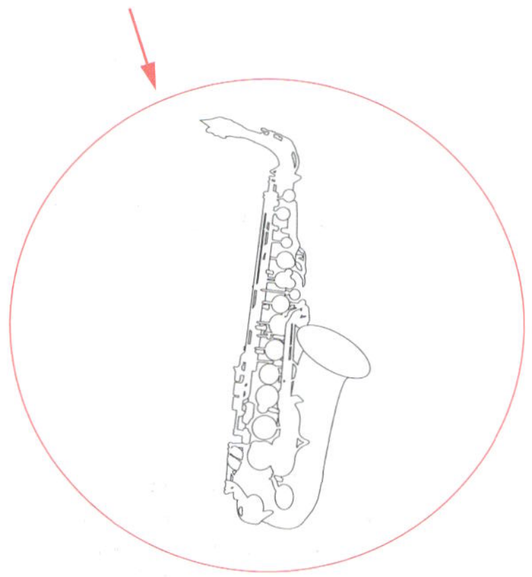

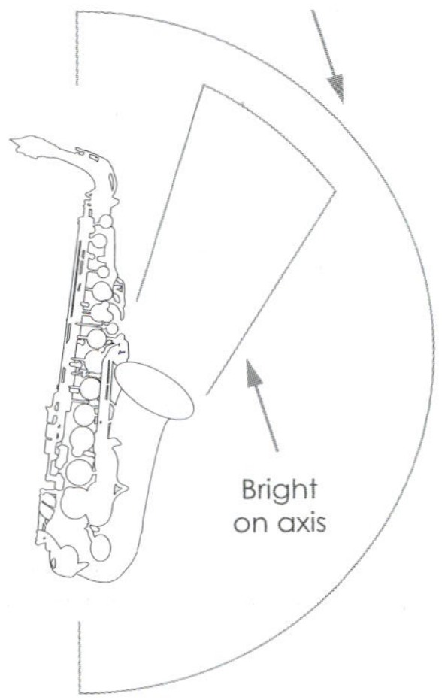

COMMON RECORDING TECHNIQUES

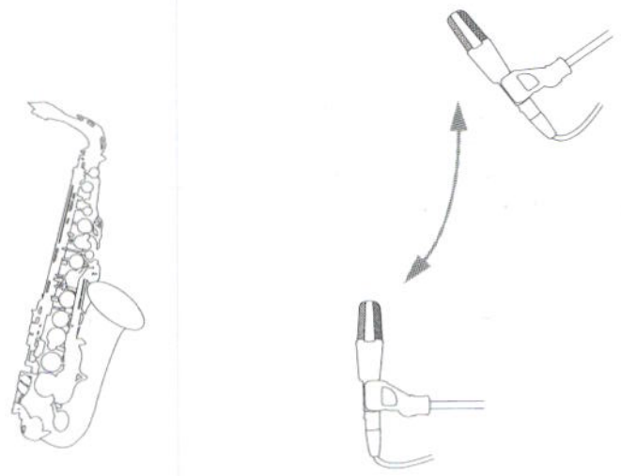

Commonly recorded from the front, slightly off-centre with the bell or on-axis with the latter, at a distance.

SOUND RADIATION

Below 311 Hz
Omni radiation

Above 311 Hz
Directive
towards the front

Bright
on axis

WOODWINDS

BARITONE
SAX

NOTES

The baritone saxophone:

- May produce high sound pressure levels (in excess of 100 dBSPL at close proximity).
- May sound overtly aggressive on the bell axis.

COMMON RECORDING TECHNIQUES

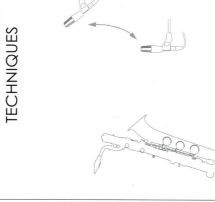

Commonly recorded from the front, slightly off-centre with the bell, or on-axis with the latter, at a distance.

SOUND RADIATION

LF
omni radiation
(below ~233 Hz)

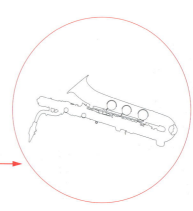

HMF / HF
bias

MF
directive
towards the front

WOODWINDS
BASSOON

NOTES

The bassoon:

- Produces low sound pressure levels at the extreme ends of its frequency range.
- Displays a dynamic range that is greatly influenced by the speed of the playing, e.g. slow passages allow for a wider range.

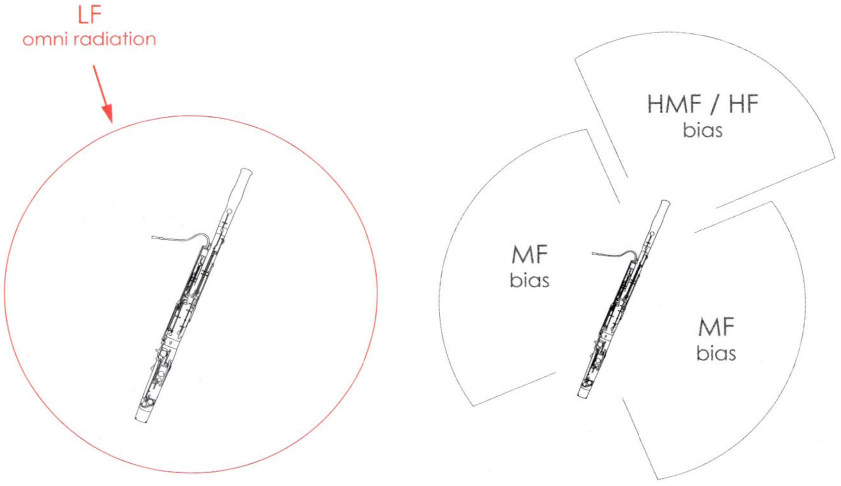

COMMON RECORDING TECHNIQUES

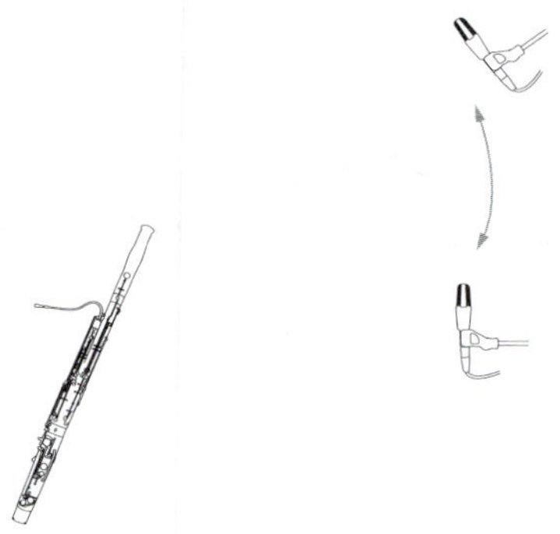

Commonly recorded on-axis, above the line of the crook or slightly off-centre with the bell, at a distance.

SOUND RADIATION

LF
omni radiation

HMF / HF
bias

MF
bias

MF
bias

WOODWINDS
Bb CLARINET

NOTES

The clarinet:

- May produce high sound pressure levels (in excess of 100 dBSPL at close proximity).
- Displays a wide dynamic range (in excess of 40 dB).

COMMON RECORDING TECHNIQUES

Reflective surface

Commonly recorded from the front, above the bell and at a distance (combining direct and reflected sound).

SOUND RADIATION

LF
omni radiation

MF
bias

HMF / HF
bias

WOODWINDS
FLUTE

NOTES

The flute:

- May sound overtly aggressive at close proximity.
- Produces low sound pressure levels at the low end of its frequency range.
- Displays a narrow dynamic range.

SOUND RADIATION

No true omnidirectional range.

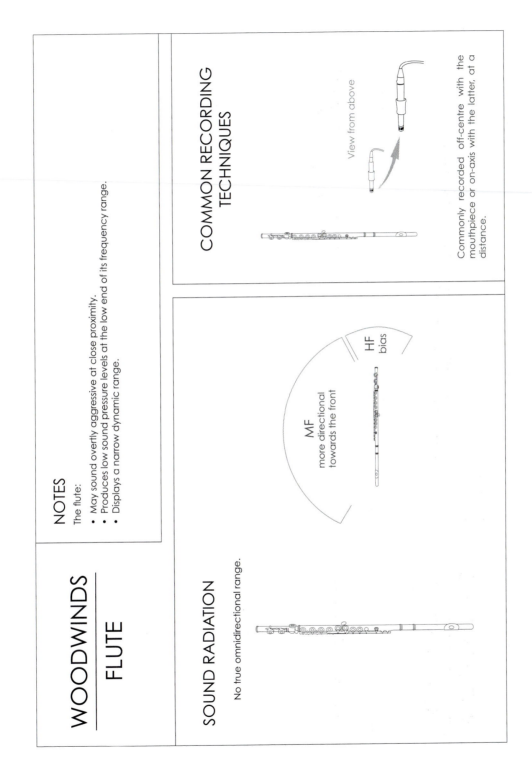

MF
more directional
towards the front

HF
bias

COMMON RECORDING TECHNIQUES

View from above

Commonly recorded off-centre with the mouthpiece or on-axis with the latter, at a distance.

WOODWINDS
OBOE

NOTES

The oboe:

- May produce high sound pressure levels (in excess of 100 dBSPL at close proximity).
- Produces lower sound pressure levels at the bottom end of its frequency range.
- Displays a narrow dynamic range (particularly in the upper register).

COMMON RECORDING TECHNIQUES

Reflective surface

Commonly recorded from the front, above the bell and at a distance (combining direct and reflected sound).

SOUND RADIATION

LF / LMF
omni radiation
(below Bb4 approx.)

HMF / HF
bias

WOODWINDS

SOPRANO SAX

NOTES

The soprano saxophone:

- May produce high sound pressure levels (in excess of 100 dBSPL in close proximity).
- May sound overtly aggressive on axis.

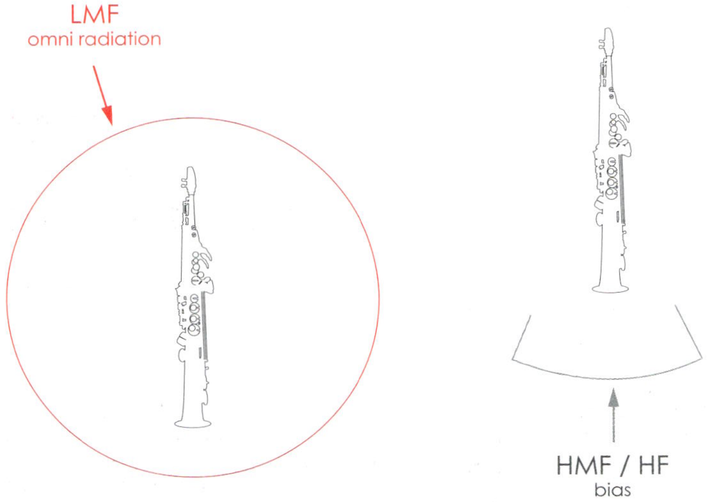

COMMON RECORDING TECHNIQUES

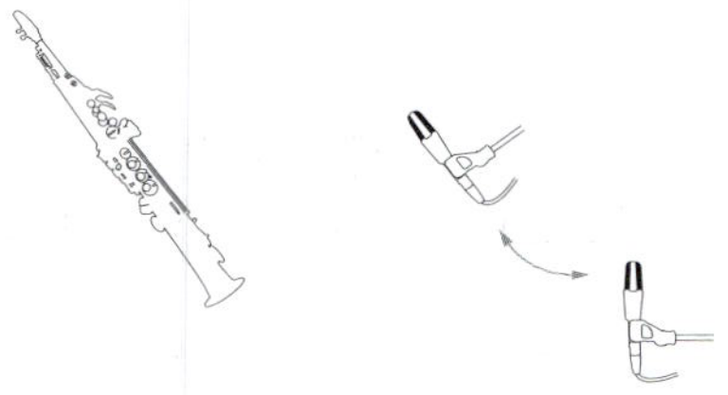

Commonly recorded from the front, off-axis with the bell.

SOUND RADIATION

LMF
omni radiation

HMF / HF
bias

WOODWINDS
TENOR SAX

NOTES

The tenor saxophone:

- May produce high sound pressure levels (in excess of 100 dBSPL at close proximity).
- May sound overtly aggressive on the bell axis.

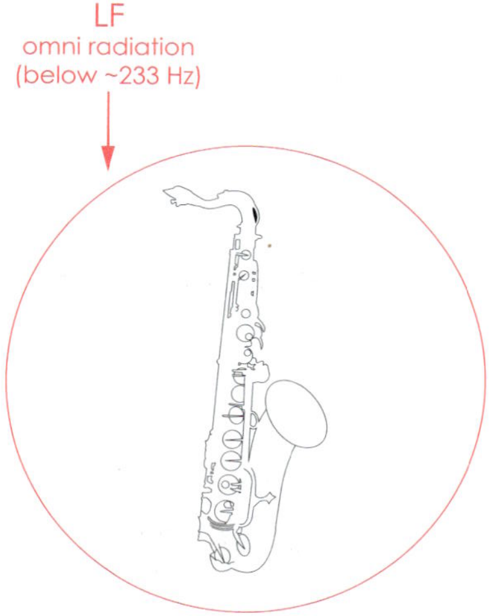

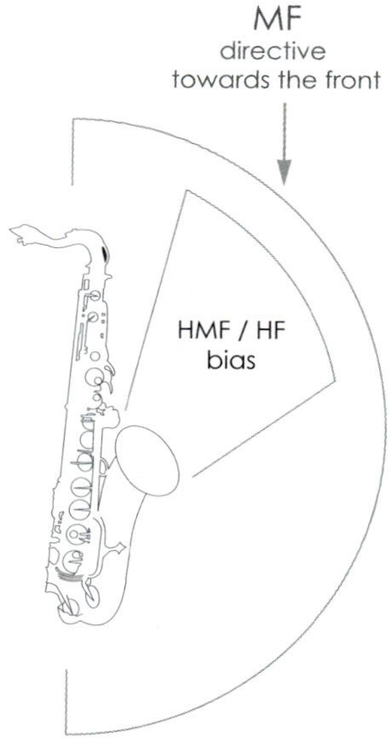

COMMON RECORDING TECHNIQUES

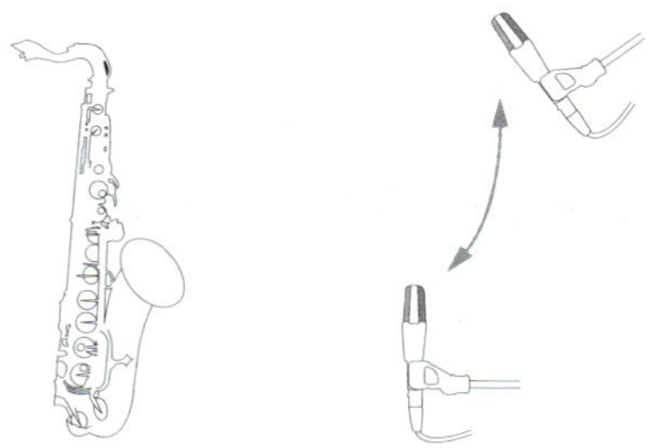

Commonly recorded from the front, slightly off-centre with the bell, or on-axis with the latter, at a distance.

SOUND RADIATION

LF
omni radiation
(below ~233 Hz)

MF
directive
towards the front

HMF / HF
bias

BRASS

FRENCH HORN

NOTES

The French horn:

- May sound unnatural at close proximity.
- Produces lower sound pressure levels at the low end of its frequency range.
- Displays a wide dynamic range.

COMMON RECORDING TECHNIQUES

Reflective surface

Commonly recorded from the front of the performer, above the instrument and at a distance (emphasis on reflected sound).

SOUND RADIATION

No true omnidirectional range.

More directive towards the rear and to the right (performer's perspective)

BRASS

TROMBONE

NOTES

The trombone:

- May produce high sound pressure levels (in excess of 100 dBSPL at close proximity).
- May sound overtly aggressive on the bell axis.
- Displays a narrow dynamic range.

COMMON RECORDING TECHNIQUES

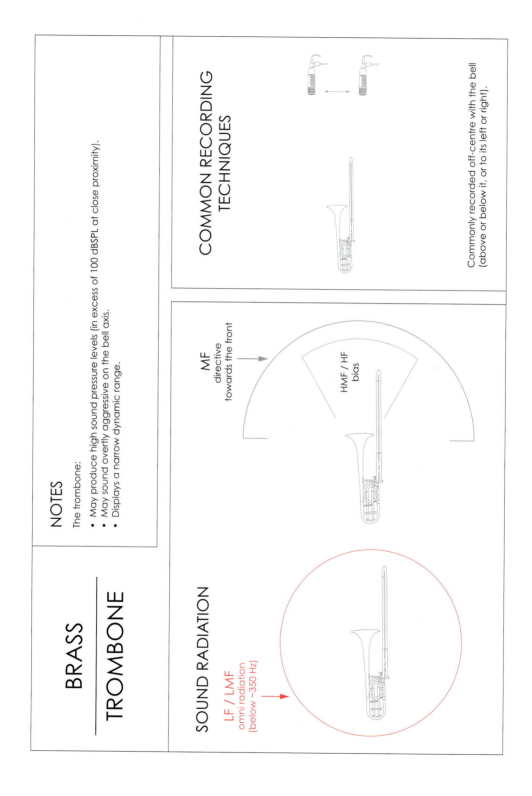

Commonly recorded off-centre with the bell (above or below it, or to its left or right).

SOUND RADIATION

LF / LMF
omni radiation
(below ~350 Hz)

MF
directive
towards the front

HMF / HF
bias

BRASS

TRUMPET

NOTES

The trumpet:

- May produce high sound pressure levels (in excess of 110 dBSPL at close proximity).
- May sound overtly aggressive on the bell axis.
- Displays a wide dynamic range.

SOUND RADIATION

LF / LMF
omni radiation
(below ~450 Hz)

MF
directive
towards the front

HMF / HF
bias

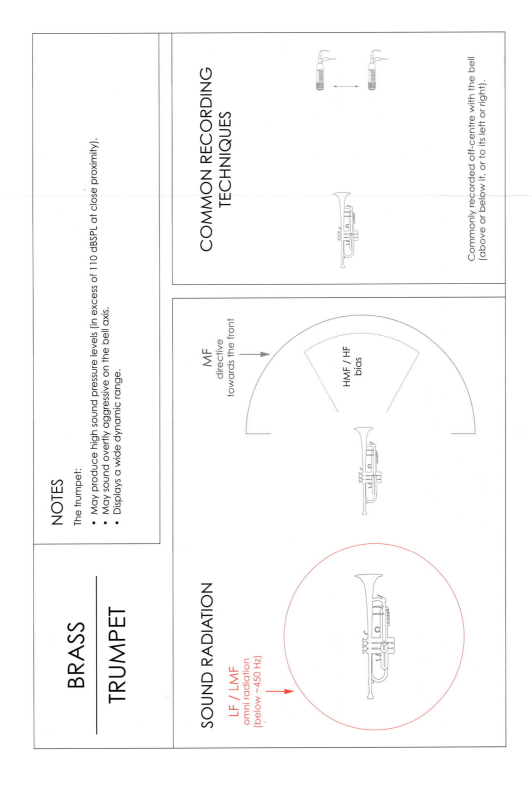

COMMON RECORDING TECHNIQUES

Commonly recorded off-centre with the bell (above or below it, or to its left or right).

BRASS

TUBA

NOTES

The tuba:

- May sound overtly aggressive on the bell axis.
- Displays a narrow dynamic range.

COMMON RECORDING TECHNIQUES

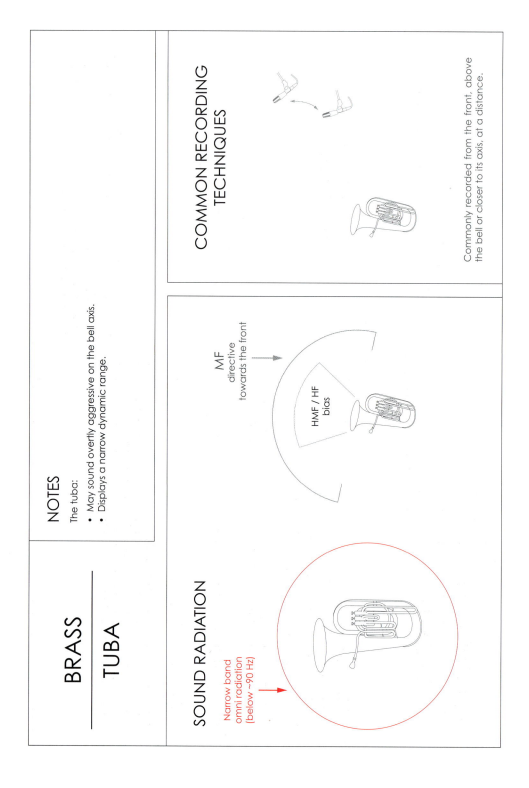

Commonly recorded from the front, above the bell or closer to its axis, at a distance.

SOUND RADIATION

MF
directive
towards the front

HMF / HF
bias

Narrow band
omni radiation
(below ~90 Hz)

PERCUSSION

BASS DRUM

GENERAL NOTES

The bass drum:

- Produces low end according to its size / depth (direct proportionality) and head tension.
- May have one (batter) or two (batter and resonant) heads.
- Is commonly played with a pedal-activated wooden (brighter) or felt-covered (duller) beater.
- Is customarily tuned between C1 and C2 .

COMMON RECORDING TECHNIQUES

Commonly recorded with one or more of the following:

- A resilient dynamic or PZM microphone placed inside the shell, slightly off-centre with the beater.
- A condenser microphone positioned at a distance (possibly angled off-axis with the instrument and/or inside a 'tunnel').
- A speaker driver at the front of the instrument.

RECORDING NOTES

- Single-headed bass drums (no resonant head) offer extra recording options, as microphones may be placed inside the shell. Such drums commonly produce a 'drier' sound with emphasis on the attack portion of the envelope, i.e. they have less resonance.

- Double-headed bass drums may require the use of multiple microphones, placed on both sides of the drum (check the polarity of the signals).

- Bass drums with a hole in the resonant head offer a good compromise between attack and resonance.

- A heavy weight, e.g. a brick, may be placed inside a bass drum to reduce unwanted low frequency 'rumble'.

- Bass drum heads may need muffling to reduce the level of undesirable harmonics or extreme resonance. A pillow may be used for such purpose, although ideally the muffler should move away from (bounce off) the head during the initial portion of the envelope. Commercial velcro pads are available for such purpose.

PERCUSSION

CYMBALS

GENERAL NOTES

Cymbals:

- Display a directly proportinal relationship between their thickness and amplitude.
- Are commonly played with wooden sticks with nylon ('brighter') or wooden ('darker') tips.
- May also be played with mallets, brushes, rakes, 'dreadlocks', rutes, etc.

COMMON RECORDING TECHNIQUES

Commonly recorded with a condenser microphone aiming down at the instrument or pointing away from the rest of the kit.

RECORDING NOTES

- The use of a dedicated spot microphone to record cymbals may not be necessary, as a drum kit overhead pair will most likely capture their sound adequately.

- A spot microphone may play an important role for the recording of cymbals when these are showcased and played lightly, e.g. with mallets.

- Microphones aimed at the edge of a cymbal will pick up a 'thinner', brighter sound, while those pointing towards the centre of the instrument will pick up a 'heavier', 'bell-like' sound.

- Pop shields may be employed for the recording of cymbals, minimising the effect of wind impact.

- Consider using a mismatched pair of sticks for the recording of cymbals, e.g. nylon tip for cymbals and wooden tip for snare or vice-versa.

PERCUSSION

HI-HAT

GENERAL NOTES

The hi-hat:

- Displays a directly proportinal relationship between its thickness and amplitude.
- Is commonly played with wooden sticks with nylon ('brighter') or wooden ('darker') tips.
- May also be played with brushes, rakes, 'dreadlocks', rutes, etc.

RECORDING NOTES

- The use of a dedicated spot microphone to record the hi-hat may not be necessary, as a drum kit overhead pair will most likely capture the sound of the instrument adequately.

- A spot microphone may play a vital role for the recording of the hi-hat in cases when the instrument is played lightly or with brushes, dreadlocks, etc.

- Microphones aimed at the edge of the hi-hat will pick up a 'thinner', brighter sound, while those pointing towards the centre of the instrument will pick up a 'heavier', more 'bell-like' sound.

- Pop shields may be employed for the recording of hi-hats, minimising the effect of wind impact.

- Consider using mismatched pairs of sticks for the playing of the hi-hat and the snare drum (in order to achieve contrasting textures).

COMMON RECORDING TECHNIQUES

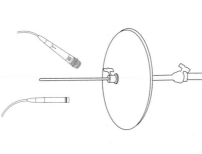

Commonly recorded with a dynamic or condenser microphone aiming down at the instrument or pointing away from the rest of the kit.

215

PERCUSSION
SNARE DRUM

GENERAL NOTES

The snare drum:

- Produces low end according to its size / depth (direct proportionality) and head tension.
- Is commonly played with wooden sticks with nylon ('brighter') or wooden ('darker') tips.
- May also be played with brushes, rakes, 'dreadlocks', rutes, mallets, etc.

COMMON RECORDING TECHNIQUES

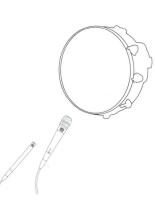

Commonly recorded with one or more of the following:

- A resilient dynamic microphone placed just outside the instrument or immediately over the rim, aiming at the point where the stick hits the head.
- A condenser microphone placed at a greater distance.
- A microphone positioned under the instrument, aiming at the snare strands (check for polarity problems - consider inverting).

RECORDING NOTES

- The snare drum may sound unnatural if recorded from too close (producing what could be described as a 'timbale-like' sound).

- High pitched bottom heads may resonate exaggeratedly.

- The number of snare strands affects the sound of the instrument considerably (use a greater number of strands for a 'John Bonham-like' sound).

- Some snare drums may resonate too much for a given application and require dampening. This may be achieved through the use of specialised materials e.g. moongel, drumgum, etc. or other substitutes such as gaffa tape, etc. (affixed close to the rim). In extreme cases, thicker options may be considered, such as tea towels, wallets, cigarette packs, etc. (which may lead to an older, 'vintage-like' sound).

- Snare drums may resonate excessively when other components of the drum kit are played, e.g. tom-toms. Dampening and changes in the tension of the bottom lugs may help reduce this problem.

- Microphones aimed at the edge of the snare drum will pick up a 'thinner' sound with a more pronounced 'ring', while those pointing towards the centre of the drum will pick up a fuller, more aggressive sound.

PERCUSSION

TOM-TOMS

GENERAL NOTES

Tom-toms:

- Produce low end according to their size / depth (direct proportionality) and head tension.
- May have one (batter) or two (batter and resonant) heads.
- Are commonly played with wooden sticks with nylon ('brighter') or wooden ('darker' sounding) tips.
- May also be played with brushes, rakes, 'dreadlocks', rutes, mallets, etc.
- Are customarily tuned at musical intervals, e.g. fourths.

RECORDING NOTES

- Single-headed tom-toms may be recorded from inside the drum shell.

- Double-headed tom-toms may be recorded from both sides of the instrument.

- Some tom-toms may require damping due to excessive resonance. This may be achieved through the same means described in the snare drum section.

- Tom-toms may resonate excessively when other components of the drum kit, e.g. the bass drum, are played. Changes in the tuning may help reduce this problem.

- Microphones aimed at the edge of tom-toms will pick up sound with a more pronounced 'ring' or resonance, while those pointing towards the centre of the drum will pick up a more aggressive sound with emphasis on the attack portion of the envelope.

COMMON RECORDING TECHNIQUES

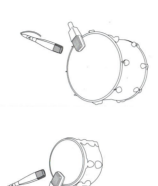

Commonly recorded with one or more of the following:

- A dynamic microphone placed just outside the instrument or immediately over the rim.
- A condenser microphone placed similarly or at a greater distance.
- A microphone positioned inside or under the instrument (consider inverting the polarity of the signal).

DRUM KIT

MONO OVERHEAD MICROPHONE

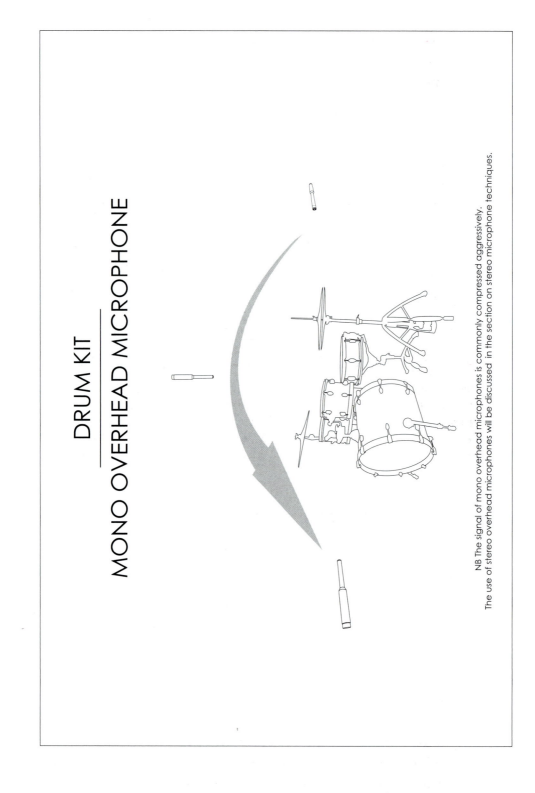

NB The signal of mono overhead microphones is commonly compressed aggressively.
The use of stereo overhead microphones will be discussed in the section on stereo microphone techniques.

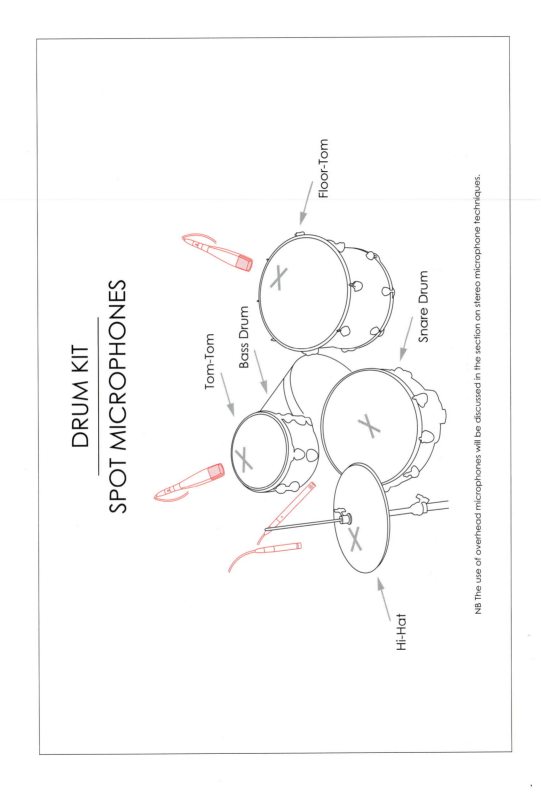

DRUM KIT
SPOT MICROPHONES

Floor-Tom

Tom-Tom

Bass Drum

Snare Drum

Hi-Hat

NB The use of overhead microphones will be discussed in the section on stereo microphone techniques.

PERCUSSION

BONGOS

GENERAL NOTES

Bongos:

- Produce high end according to head tension.
- Are commonly played by hand.
- Are customarily tuned at a musical interval, e.g. a fourth.

RECORDING NOTES

- Bongos are comprised of two drums, known as 'macho' (larger) and 'hembra' (smaller).

- Bongos may be recorded with a stereo microphone technique (explored in an upcoming section of this book).

- Bongos may require recording at close proximity (if played lightly).

- Bongos sound best in controlled reflective environments, i.e. rooms with diffused reverberation that is not too bright.

- Bongos may be recorded on stands or between the player's legs (placing bongos directly on the floor should be avoided).

COMMON RECORDING TECHNIQUES

Commonly recorded with one or more of the following:

- A microphone placed just outside each of the two drums or immediately over their rims, aiming at the point where the fingers meet the head.
- A microphone (possibly stereo), centred with the drums.
- A matched stereo pair of microphones.

PERCUSSION

CONGA

GENERAL NOTES

Congas:

- Produce low end according to their size / volume (direct proportionality).
- Produce high end according to head tension.
- Are commonly played by hand, although they may also be played with very light sticks.
- Are usually tuned to a musical interval.

RECORDING NOTES

- The term 'congas' is loosely applied to a number of hand drums known individually as 'tumbadora' (the largest), 'conga' (mid-size) and 'quinto' (the smaller instrument).

- A fourth, smallest soloing drum known as 'requinto' is sometimes added to a set of congas.

- Congas record well in large rooms.

- Congas sound best in controlled reflective environments, i.e. rooms with diffused reverberation that is not too bright.

- Congas may sound unnatural if recorded at close proximity.

- Congas may be recorded on stands (less low end) or directly on the floor (more low end).

COMMON RECORDING TECHNIQUES

Commonly recorded with one or more of the following:
- A microphone placed outside the drum or close to the rim, aiming at the point where the player's hand meets the head.
- A stereo microphone or matched stereo pair (when multiple congas are used).

PERCUSSION

DJEMBE, DOUMBEK, ETC.

GENERAL NOTES

Djembes, doumbeks and other similarly shaped hand drums:

- Produce low end according to their size / volume (direct proportionality).
- Produce high end according to head tension.
- Are commonly played by hand.

RECORDING NOTES

- Djembes and similarly shaped hand drums may offer lug or rope-based tuning systems.

- Such drums may require the periodic application of heat for tuning.

- Multiple microphones may be required for the recording of the djembe, doumbek, etc. (to capture their low/high end).

- Such drums record well in large rooms.

- Djembes and similarly shaped drums sound best in controlled reflective environments, i.e. rooms with diffused reverberation that is not too bright.

- Djembes may sound unnatural if recorded at close proximity.

- Doumbeks may require close microphone positioning (if played with fingers, i.e. quietly).

- Such drums may be recorded on stands or held / hanging from a player's body.

COMMON RECORDING TECHNIQUES

Commonly recorded with one or more of the following:
- A microphone placed in front of the drum, at a distance.
- A microphone positioned behind the drum (capturing low end only).

VOICE

NOTES

The human voice:

• May produce high sound pressure levels (in excess of 100 dBSPL at close proximity).
• Has an extended dynamic range (commonly recorded with compression).
• Is frequently recorded with a 'pop shield', employed to reduce the effects of plosives and sibilance.

COMMON RECORDING TECHNIQUES

Commonly recorded from the front.

SOUND RADIATION

Directive towards the front

HMF / HF bias

No true omnidirectional range.

Extra Microphones

Recordists aiming to be prepared for common session eventualities may setup a few 'standby' microphones, placing them slightly outside the immediate recording area. These extra transducers may be brought out and connected rapidly in the case of new requirements, e.g. spontaneous instrumentation ideas, etc.

Using Multiple Microphones to Create Texture

In some circumstances, the use of a single microphone may not be sufficient to generate a full, well-balanced picture of a sound source. In such cases, it is not uncommon for engineers to use multiple 'mics' to capture the different sound attributes of an instrument and/or the acoustic qualities of the recording environment. This approach may help minimise or eliminate the need for processing (equalisation) during mixdown, as engineers might use the blending microphone signals at varying levels (but at a common 'pan' position) to create different textures, e.g. brighter, darker, etc.

It is important to remember that if two microphones are placed at different distances in relation to a sound source, their combined signal may present polarity problems (comb filtering). If the two signals interfere with each other negatively (and this must be evaluated by ear), recordists may need to use signal processing (delay) to ensure that they will not have a time gap between them.

The aforementioned use of multiple transducers must not be mistaken for a stereo microphone technique, as in the latter, two or more microphones are used to establish localisation and their individual signals are commonly panned *separately* across the stereo field.

Using Delay to Avoid Comb Filtering

The example below illustrates how a delay unit may be used to remedy the (possibly) negative comb-filtering effects resulting from the sum of signals from microphones placed at different distances from a sound source. The delay value is chosen to ensure that both signals will reach the recorder at the same time, i.e. the signal from

Using Delay to Avoid Comb Filtering (continued)

the microphone closest to the amp is delayed by the amount of time that it takes sound to cover the extra metre between the two transducers. The equation presented here uses 344 metres per second as the speed of sound (corresponding to a temperature of approximately 21˚C).

Calculating Delay Time

$$c = \frac{d}{t}$$

$$t = \frac{1}{344}$$

$$t = 0.0029 \text{ seconds}$$

Where:

c = speed of sound (344 m/s)

d = distance between 'mics' (1 m)

t = delay value needed

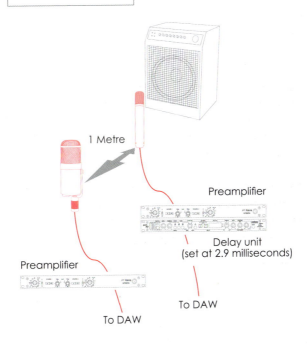

1 Metre

Preamplifier

Delay unit
(set at 2.9 milliseconds)

Preamplifier

To DAW

To DAW

DAW users may alternatively record the signal from the first microphone without delay and 'nudge' its corresponding region by 2.9 milliseconds or by 128 samples (if recording at 44.1 kHz).

STEREO MICROPHONE TECHNIQUES

Stereo microphone techniques may be defined as the simultaneous use of two or more microphones to record a common sound source, as long as the signals they produce are unique and, more importantly, that they are panned opposite across the stereo field during playback or reproduction. Such techniques are commonly used to record ensembles, e.g. orchestras, string quartets, etc, and instruments such as the piano, vibraphone, glockenspiel, marimba, drums, etc.

Stereo microphone techniques are broadly categorised according to the physical distance between transducer diaphragms:

- **XY / Coincident Techniques** are those in which microphone capsules are placed immediately over one another and are set at different angles in relation to the sound source.

- **Near-Coincident Techniques** are those in which microphones are placed a few centimetres from one another and are set at different angles in relation to the sound source.

- **AB / Spaced Pair Techniques** are those in which microphones are placed more than (approximately) 30 centimetres apart. NB the term AB is used in some European countries to describe coincident pair techniques.

Three main factors influence stereo imaging when multiple transducers are used in tandem to record a common sound source. They are:

- Amplitude differences
- Time differences
- Spectral differences.

The aforementioned factors also dictate localisation in human hearing and it is important for recordists to understand their influence over the efficiency of a microphone technique. When the same sound source is picked up by two matched transducers used in a stereo configuration, localisation during reproduction is determined by:

- How much greater in amplitude the output of one transducer is in relation to the other (an approximate 15 dB difference implies a full left or right reproduction if the two microphone signals are panned 'hard-left' and 'hard-right').

- How much earlier one transducer picks up the source sound in relation to the other (an approximate 1.5 millisecond difference implies a full left or right reproduction if the two microphone signals are panned 'hard-left' and 'hard-right').

- (To a minimal degree) how much more harmonically rich the output of one transducer is in relation to the other (in relation to the same source).

A combination of the differences described above will be present when multiple microphones, used to record a common sound source, are set at different distances and angles in relation to each other.

XY / Coincident Pair Techniques

In XY or coincident pair techniques, amplitude differences are essentially responsible for localisation, as there is no time-delay gap in sound pick-up between the two capsules (it is important to note that spectral differences may also play a minimal part in this process, if off-axis colouration is sufficient to generate considerable contrast between the two signals).

Coincident pairs:
- Are comprised of directional or bidirectional microphones
- Generate mono compatible signals (no time delay)
- Are least effective if placed distant from the sound source(s).

COINCIDENT PAIRS

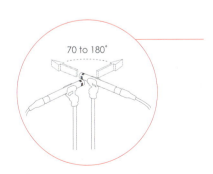

70 to 180°

Basic XY

In this technique, two matched unidirectional microphones are placed above one another (with capsules almost touching) and set at a left-to-right angle between approximately 70° and 180°.

COINCIDENT PAIRS

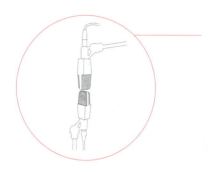

Blumlein Technique

In this technique, two matched microphones set at a figure-of-eight polar pattern are placed above one another, at an angle of 90°.

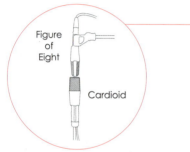

Middle and Sides

In this technique, a cardioid microphone faces the sound source directly and a second, figure-of-eight transducer is placed immediately above or below it, at a 90° angle. The signal from the bi-directional microphone is duplicated and one of its two versions has its polarity inverted. When monitoring, the 'middle' directional microphone is panned to the centre of the stereo field, while the two 'side' channels are panned 'hard-left' and 'hard-right' respectively. The amount of stereo spread is dictated by the amplitude of the 'side' signals.

Using Middle and Sides

The M&S technique allows for the remote control of stereo width through changes in the level balance between the middle and the side components, e.g. the higher the level of the sides the wider the image. This technique can be particularly effective when used in live sound applications, as engineers can quickly change the overall stereo spread of recorded and front of house signals without repositioning microphones.

The following is a depiction of the M&S set-up when a console is used:

Using Middle and Sides (continued)

Recordists must pay close attention to the spectral changes that result from the rebalancing of levels when using M&S pairs, as it is not uncommon for signals to become 'thinner' as the amplitude of the 'sides' is increased. As a solution to this problem, the use of equalisation, e.g. low-mid shelving boosts, may help maintain the integrity of the material.

Near-Coincident Pair Techniques

In near-coincident techniques, amplitude is still primarily responsible for establishing localisation, although time arrival differences start to contribute to the stereo effect (once again spectral differences may play a very minor role in the process).

Near-coincident pairs:
- Are comprised of directional microphones (unless 'baffled').
- May not be mono compatible (check).
- Provide a greater sense of spaciousness (compared to XY techniques).
- May be placed at a greater distance from the sound source.

NEAR-COINCIDENT PAIRS

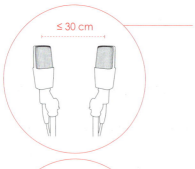

Basic Near-Coincident

In this technique, two matched unidirectional microphones are placed a few (~ 30 or less) centimetres apart and set at a left-to-right angle.

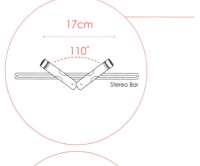

NOS

In this technique, two matched cardioid microphones are placed 30 centimetres apart and set at a 90° angle.

ORTF

In this technique, two matched cardioid microphones are placed 17 centimetres apart and set at a 110° angle.

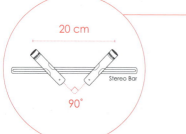

DIN

In this technique, two matched cardioid microphones are placed 20 centimetres apart and set at a 90° angle.

'Baffled' Near-Coincident Techniques

In 'baffled' near-coincident techniques, spectral differences play an important role in establishing localisation, as these techniques aim to emulate the human hearing mechanism.

The following are examples of 'baffled' near-coincident pairs:

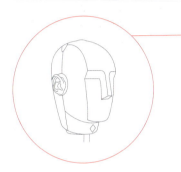

The 'Dummy' Head

'Dummy' head binaural stereo microphones are built with materials that have similar properties (mass and density) to those found in a human head. Such microphones are commonly used to capture sound on location when the aim is to provide the listener with the sensation of 'being there'. The use of 'dummy' head transducers is particularly efficient for the generation of binaural recordings to be monitored via headphones.

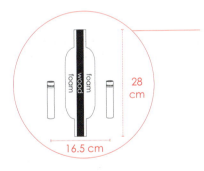

OSS / Jeklyn Disk

OSS systems are commonly referred to as 'quasi-binaural'. Here, two omnidirectional transducers are spaced 16.5 centimetres apart and are separated by a rigid disk of 28 centimetres diameter. The disk is covered with sound-absorbing material in an attempt to simulate the properties of a human head, i.e. the system is only directional at higher frequencies.

AB / Spaced Pair Techniques

In AB or spaced pair techniques, a combination of amplitude and time arrival differences are responsible for localisation.

Spaced pairs:

- Are commonly comprised of omnidirectional (less colouration) or hypocardioid microphones.
- May not be mono compatible (check).
- May generate a 'hole in the middle' during reproduction (a third, 'middle' microphone may help with this).
- Provide the greatest sense of spaciousness.
- May be placed at the greatest distance from the sound source.

In basic AB techniques, two omni (preferably) or directional microphones are positioned over 40 centimetres apart. Spaced microphone pairs are commonly used to record drums and other wide image sources, as they are very easy to set-up and reposition.

SPACED PAIRS

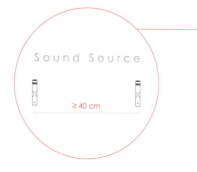

Basic AB

In basic AB techniques, two omni-directional (preferably) or directional microphones are positioned over 40 centimetres apart. Spaced microphone pairs are commonly used to record drums and other wide image sources, as they are very easy to setup and reposition.

The Decca Tree

The technique known as 'Decca Tree' has had many variants through the years, although it may be generally described as an arrangement of three omnidirectional microphones mounted on a 'T' shaped bar, where the centre transducer is placed 1 to 1.5 metres ahead of the outriggers, which in turn are spaced by approximately 2 metres. It is common practice to use Decca Trees for the recording of orchestras, where the array is positioned a metre or so behind and 2 to 3 metres above the conductor's position.

Choosing A Stereo Microphone Technique (An Ear-Based Approach)
The following is a suggested strategy for the selection of stereo microphone techniques for the recording of a small ensemble or instrument section:

- Approach the sound source and determine its 'centre' axis, i.e. an imaginary line that seems to separate its low and high registers or the axis of symmetry regarding the energy produced.

- Move away slowly (following the axis), listening for spectrum changes and for the balance between 'dry' and reverberant sound.

- Determine the point(s) where low and high frequencies are balanced, reverberation is optimum and unwanted sounds are low in amplitude (if the points do not coincide find a compromise placing emphasis on spectral balance as artificial reverberation may be added later).

- Check how much of the original stereo image is still preserved at the selected position.

- Base the choice of microphone technique (including polar patterns) on how much separation is still detected at the chosen position and on the desired spread during reproduction. As general rules-of-thumb, under normal circumstances:

  1. Directional polar patterns and 'tighter' arrays are commonly effective in closer proximity to 'stereo' sound sources.

  2. The distance between microphones may be seen as directly proportional to their distance to the sound source, i.e. the further the distance from the sound source the more spaced the capsules should be (some recordists use a 3:1 rule for spaced pairs, where every metre from the sound emitter equates to 3 metres between transducers).

- Audition the sum of the microphone signals in mono and check for polarity problems, e.g. invert the polarity of one of them.

Choosing A Stereo Microphone Technique (continued)

- Audition the signals from the microphones panned 'hard-left' and 'hard-right' respectively and check for the localisation and the focus of each of the individual sound source components.

- Do not forget the possibility of requesting a change in spatial arrangement for small groups or ensembles, e.g. XY pairs may be best at closer proximity to sound sources configured in an arch formation, while near-coincident pairs may be best at closer proximity to sources placed on a straight line.

Stereo Microphone Techniques and Orchestral Recordings
Orchestras and large groups are commonly recorded with a combination of stereo pairs and 'spot' microphones. The following is one of many possible approaches:

- Set up a main coincident or near-coincident pair behind the conductor, at enough height to avoid possible acoustic shadowing (a simple way to determine a suitable height is to check if the microphone has a 'line of sight' with all the musicians).

- Set up a spaced pair of omnidirectional microphones '(outriggers'), to the left and right of the original pair (consider lining them up with the centre of each of the two halves of the ensemble respectively).

- Audition the first (main) pair in mono and check for polarity problems (cancellations or 'holes' in the spectrum).

- Audition the same pair panned 'hard-left' and 'hard-right' respectively and analyse the localisation and the focus of each of the individual sound source components, e.g. first and second violins, celli, basses, etc. (check the 'phase' or polarity using a metering device, if available).

Stereo Microphone Techniques and Orchestral Recordings (cont.)

- Repeat the two previous steps for the 'outriggers' in isolation and add their panned signals to the mix, checking that they do not 'smear' or fight the stereo image established by the main pair.

- Add 'spot' microphones to reinforce the sound of sections, e.g. woodwind, brass, etc. or single instruments that are not picked-up by the two pairs at a sufficient level, e.g. celesta, harp, etc.

The Stereo or Stereophonic Zoom

Engineer and author Michael Williams has coined the term 'Stereophonic Zoom' to describe a system that facilitates the selection of the most appropriate stereo microphone technique for unique recording conditions, i.e. on a case-by-case basis. When using the aforementioned system, recordists can choose their desired sound reproduction localisation characteristics, while minimising angular distortion or the 'stretching' of images. Engineers that are employed for the high-fidelity recording of acoustic ensembles or instrumentalists must familiarise themselves with this arguably definitive treatise on stereo microphone techniques.

The following pages contain examples of stereo pair applications.

STEREO MICROPHONE TECHNIQUES

DRUM KIT

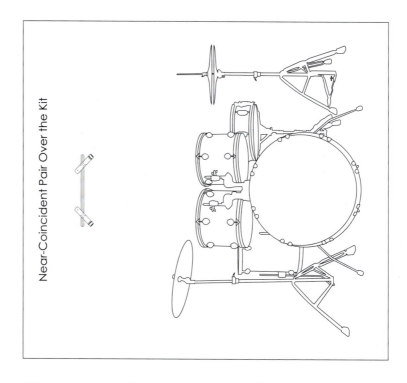

Coincident Pair Over the Kit

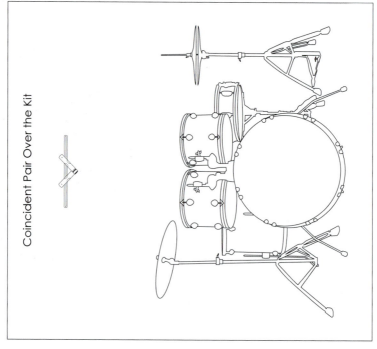

Near-Coincident Pair Over the Kit

STEREO MICROPHONE TECHNIQUES

DRUM KIT

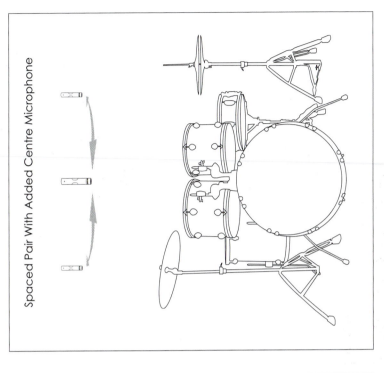

Spaced Pair With Added Centre Microphone

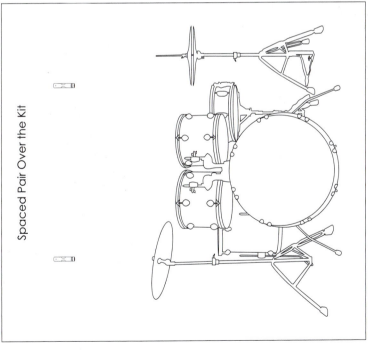

Spaced Pair Over the Kit

STEREO MICROPHONE TECHNIQUES
DRUM KIT (USING POLARITY INVERSION)

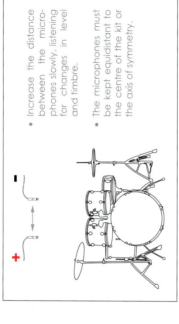

- Place two overhead microphones at very close pproximity.
- Assign each microphone to a separate channel.
- Pan the two signals to the centre and invert the polarity of one of the channels (the overall sound should be very quiet and thin).

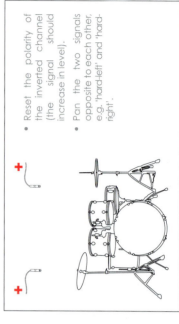

- Increase the distance between the microphones slowly, listening for changes in level and timbre.
- The microphones must be kept equidistant to the centre of the kit or the axis of symmetry.

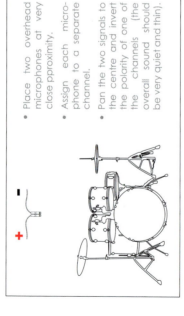

- At a given distance the combined signals should resemble the initial sound (quiet and thin).
- Stop moving the microphones

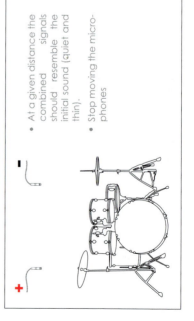

- Reset the polarity of the inverted channel (the signal should increase in level).
- Pan the two signals opposite to each other, e.g. 'hard-left' and 'hard-right'.

STEREO MICROPHONE TECHNIQUES

DRUM KIT

SPACED PAIR WITH CENTRED SNARE AND BASS DRUM*

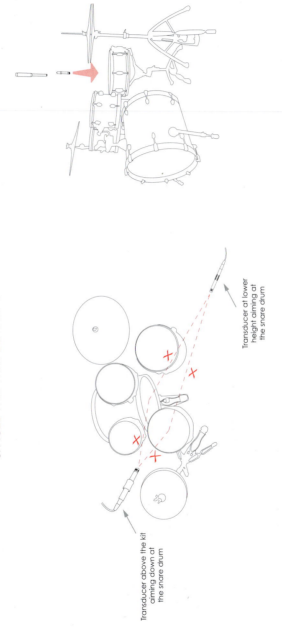

Transducer above the kit aiming down at the snare drum

Transducer at lower height aiming at the snare drum

Both transducers are at the same distance **X** from the snare and the bass drum.

* This microphone setup with an added spot microphone for the bass drum is also known as the Glyn Johns technique.

STEREO MICROPHONE TECHNIQUES

DRUM KIT

SPACED PAIR BEHIND THE KIT

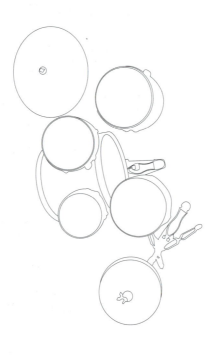

Microphones set slightly above the drummer's head, pointing at the kit.

This microphone array is sometimes referred to as a 'Nashville Pair', although the reason for such denomination is unclear.

STEREO MICROPHONE TECHNIQUES

GRAND PIANO

COINCIDENT PAIRS OUTSIDE THE INSTRUMENT

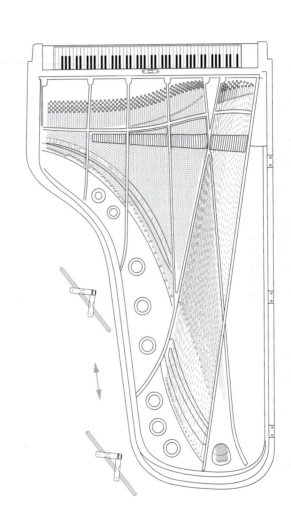

Consider moving the pair in order to achieve a balance between attack (hammers) and body (wood).

STEREO MICROPHONE TECHNIQUES
GRAND PIANO

SPACED PAIRS

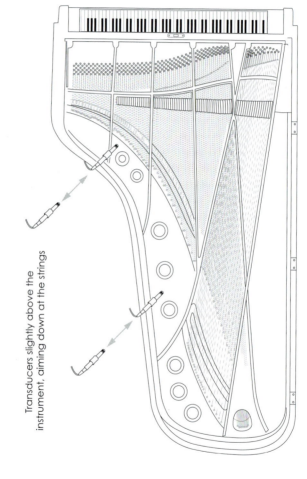

Transducers slightly above the instrument, aiming down at the strings

Consider moving the pair into the piano in busy productions where the instrument might not cut through the mix.

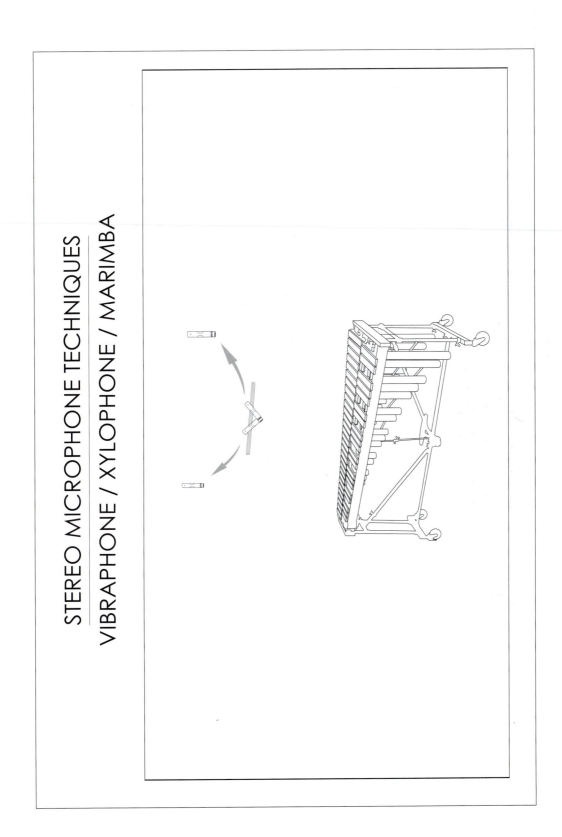

STEREO MICROPHONE TECHNIQUES
VIBRAPHONE / XYLOPHONE / MARIMBA

THE CONTROL ROOM SET-UP

As one or more members of the recording team setup the performance space or live room for a session, someone (commonly the main engineer) should be responsible for the preparing and running of the equipment in the control room, i.e. although on occasion individuals may be required to run sessions on their own, approaching all tasks sequentially, under ideal circumstances at least two people should work simultaneously for the purpose of tracking.

As we have seen, certain procedures can be carried out well in advance to the arrival of the musicians to the studio, e.g. testing, calibration, DAW file creation, etc., so when the performers are ready to play, recordists should be fully prepared for sound check. This should help maintain a creative and energetic atmosphere, commonly referred to as the 'vibe', which should be seen as the most important element in any session (it is not possible to overemphasise this last statement). A good production team is able to maintain a positive mood in the studio, through the minimising of the time gap between set-up and recording and through the creation of a psychologically motivating environment for those who are waiting.

Setting DAW Preferences

Most DAWs allow the user to select their desired hardware buffer size. This set-up step is very important, as it may have a significant impact on latency and therefore the success of a session. As a general rule for recording, engineers must aim to keep the buffer size at a low number of samples to ensure that signals can reach the DAW and leave to be subsequently monitored with a minimum of delay. During mixing, buffer sizes should normally be increased in order to allow the DAW to manipulate and sum signals with no insufficient processing power related problems (here latency is not an issue).

Music Playback During Set-Up

Some recordists/producers may find it beneficial to play music during set-up, as this may:
* Help the team tune their ears to the recording environment
* Inspire the artists
* Remind all involved of why they are there.

GAIN STRUCTURE

Gain structure has an enormous impact on the quality of recordings and it is one of the main aspects separating professional work from amateur attempts. In broad terms, levels should be set so signals conform to the dynamic range of the audio chain and, more specifically, to that of the recording medium, i.e. gain should be set high enough for the material to be well above the chain's noise floor at quiet passages, while staying below the clipping point during the loudest ones. This appears simple enough until one considers how often levels may be manipulated between the first input stage of a signal path and the output to the recorder. In the case of analogue consoles, this could include the gain of a microphone preamplifier and of an equaliser(s), dynamics processor(s), channel fader, group buss summing amp, etc., therefore it is vital for recordists to understand the role of each of the aforementioned elements in order to optimise (or avoid) their use. As a rule of thumb, signals must be made as robust as possible at the earliest possible stages of a signal chain, as any amplification at later points will not only raise the level of the material, but also that of any noise picked up along the way.

The following is an example of the possible gain stages that a signal may go through before reaching a recorder:

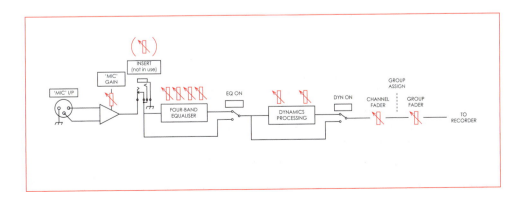

In the previous diagram we see a microphone input signal going through a preamplifier, a possible insert point, a four-band equaliser. a dynamics processing section comprised of a gate and a compressor, a channel fader and a group buss fader before reaching the recording device.

Recordists should only evaluate the suitability of signals after determining their correct gain structure. Please refer to Appendix 5 for an example of gain setting procedure.

Compression During Tracking

It is not uncommon for some sound sources to have a dynamic range that exceeds that of the recording signal chain. As an example, untrained (and in some cases even experienced) singers may have such an enormous dynamic range that if levels are set so they are adequate for the recording of quiet passages, they will lead to clipping during the loudest moments of the same recording. As a solution to such problem, engineers have long been using dynamic range processing (compression) to reduce the gap between level extremes.

Metering

The metering of signals is extremely important for both the calibration and the running of audio systems. A few different standards exist and some of them have been in use for over 70 years. The following is an overview of the more common types of meters:

VU (Volume Units – IEC60268-17)
VU meters work on level averages with a behaviour that approximates (or aims to emulate) that of the human ear. This type of metering commonly spans between – 20 and + 3 VU and presents a rise time of 300 milliseconds (the time taken for the meter to reach full deflection for a steady tone input signal).

Recordists working with VU meters must consider the implications of their slow ballistics and learn to interpret readings. The following are a few observations regarding volume units:

Metering (continued)

- Transient (short impulse) signals may fail to deflect VU meters

- Peaks for shorter impulse signals may correspond to values that are 6 to 20 decibels over what is displayed on a VU meter, e.g. when recording transient rich instruments such as drums, operators should not let the VU meter deflect much past – 5 VU.

PPM (Peak Program Meter)

Peak program meters aim to provide more instantaneous level readings. Original BBC UK PPM (IEC 268-10 IIA) devices' readouts span from 1 to 7, where 4 equates to 0 dBu and 5 to + 4 dBu. Another version of PPM meters, the EBU PPM (IEC 268-10 IIB) spans from – 12 to approximately + 14, where a 'Test' readout equates to 0 dBu and all other readings correspond their values in dBu, e.g. + 4 EBU PPM = + 4 dBu). It is important to point out that 'true' audio waveform peaks may be 6 to 8 decibels higher than what is displayed on PPMs.

EBU Digital (IEC 60268-18)

EBU digital meter readings span from – 40 (negative fourty) to 0 dBFS (the clipping point). In purely digital metering, the establishing of average operating levels is not as important as the detection of peaks.

THE SOUND-CHECK

The sound-check stage of recording sessions requires great focus from the production team and ideally only those directly involved in the process should be present in the studio. At this point any distractions can have serious consequences and delay production considerably. The following are notes that may help streamline/optimise the sound evaluation procedure. During sound-check:

- All mobile phones must be switched off and not simply set on 'silent' or 'vibrate' mode (use 'airplane' mode as an alternative).

- The recording environment should ideally be free of still and video cameras, as all efforts and attention should be dedicated to the appraisal and optimising of sound (filming commonly inhibits the production team at such an early stage).

- Talkback communication must be kept constant, with no extended gaps between performance and control room feedback.

- Musicians should not have to wait in the studio unnecessarily, i.e. producers should stagger arrival times.

- Performers must play the part that they will be recording at its appropriate level and intensity (although they may start the process by playing something else in order to avoid fatigue).

- Drummers should play their whole kit (the 'groove') and not individual elements in isolation, i.e. they should perform their part using the whole instrument, even during the evaluation of snare or bass drum sounds, etc. The rationale for this comes from the fact that drummers have a tendency to play unnaturally when hitting a single component of the kit repeatedly, presenting recordists with an unrealistic picture of what will happen later.

- Drummers should remove all unused elements from the kit, e.g. toms.

- Musicians should be allowed and, if appropriate, encouraged to play using effects processors, e.g. guitarists using effects pedals, etc., although recordists should ideally split the signal and evaluate/capture both the 'dry' (DI) and the 'wet' versions of the instrument's sound (allowing for subsequent 'reamping').

- The team should not be afraid to change instruments, strings, drum heads, microphone positions, etc. to improve sounds that do not correspond to expectations.

- Engineers should only resort to the use of equalisation if changes in microphone and/or position do not seem to yield the desired results.

Reamping and the Sound-Check

The use of a 'reamp' box may facilitate the sound-check procedure for guitarists, who may record their part and audition it played back through a few different amplifier models. This may help musicians evaluate sound objectively, as some individuals can appraise sound better (and accept criticism) when they are not playing their instruments.

Polarity Problems

It is possible for multiple microphones to present polarity problems when combined, producing a 'hollow' or 'thin' sound. As an example, two 'mics' aimed at the top and the bottom of a snare drum respectively may require the use of the polarity inverting switch of one of their corresponding channels. In situations when such a switch is not available, technicians may need to wire a polarity inverting cable, e.g. an XLR lead with pins two and three from one end connected to pins three and two of the other end respectively. NB Ensure such cables are marked very clearly. In extreme circumstances engineers may record signals that are incoherent polarity-wise and use the audio region inverting function of a DAW after recording.

Polarity coherence should ultimately be evaluated by ear during sound-check.

THE MUSICIANS' CUE MIX

It is not possible for artists to perform well if they cannot hear themselves appropriately and although this may seem obvious, some engineers do not appear to understand the importance of the cue mix. The musicians' monitoring experience while recording can affect every aspect of their performance and in controlled environments, e.g. in a studio, it should be possible (and advisable) for engineers to create monitoring mixes that are more than just suitable. When artists are provided with outstanding cue mixes, they are encouraged to perform at a high level and not uncommonly if what they hear has a 'record-like' quality they will seem more capable of delivering 'record-like' material.

Performance monitoring set-ups can vary greatly, as some artists refuse to record wearing headphones while others are happy to create their own in-ear cue mixes. Regardless of method, engineers should always check and optimise the monitoring conditions during sessions.

The playing of pre-recorded music may help engineers set up headphone or 'cue' mixes, as some musicians are happy to play along other artists' music while they select the tone for their instruments and amplifiers (although it is imperative to stop playback immediately if anyone chooses to work on their own part from the very start). The production team should be ready to start recording to the multitrack device as soon as the performers commit to the cue mix verbally (and ideally the team will have started recording to a two-track machine from the very beginning of the session).

Communication / Talkback

The establishing of communication between live and control rooms is of great priority and ideally the talkback system should be set up and checked prior to the beginning of sessions. There should be a constant dialogue between engineers and artists, as otherwise a sense of isolation may affect performance.

Singer Cue Mixes

The following are a few observations that may help recordists with their singers' monitoring set up:

- Do not be reluctant to add reverberation to the headphone mix, as this may help singers stay in tune (as they are able to listen to the 'tail' of notes).
- Be attentive to cue mix levels, as performers may sing 'flat' if bass content is monitored at very high volume through headphones (humans perceive low-frequency content at a lower pitch when auditioning at high amplitudes).
- Place singers in the control room and get them to perform with the monitors at live performance-like levels. This may help them sing more energetically and with intention (ensure to use directional dynamic microphones placed with their null points directed towards the speakers). Pay close attention to pitch.

Headphone 'Leakage'

When performers set their cue mixes at very high levels, the output of headphones may be picked up by microphones during tracking. This is referred to as headphone 'leakage' or 'bleed' and it commonly affects the recording of singers and acoustic instrumentalists. The following may help minimise this problem:

- Use single-sided headphones whenever musicians express the desire to keep one ear free to listen to the acoustic sound in the live room.
- Disconnect unused headphones, i.e. mute the headphones of musicians that are not playing on an overdub and are leaving the live room.
- Use closed back headphones.
- Use EQ to minimise the audibility of unwanted signals such as a click-track (or use a 'darker' sounding click).

The Use of Groups for the Creation of Cue Mixes

The use of group busses can help engineers create easily modifiable cue mixes, as less modules or controls may be required for the alteration of levels, e.g. by using a stereo drum group, engineers may raise the level of all kit components in the cue mix simultaneously by using two auxiliary send controls (as opposed to having to modify multiple individual component levels to the cue mix auxiliary busses).

TRACKING

When a project finally reaches the recording stage after days or possibly weeks of planning and preparations, it should become clear to the team if their pre-production time was spent wisely. At this point, professionals with strong organisational skills should be able to perform their tasks with little or no surprises.

As the multitrack device starts recording, the production team must be working at their highest level of concentration. At this point, having multiple pairs of ears and eyes in the studio is an invaluable asset, as this allows engineers to dedicate their attention to the running of equipment while producers evaluate the quality of performances, etc.

Continuous Two-Track Recording

Ideally, a two-track recorder should be set to record everything that happens in a session, i.e. a recording device should run from the sound-check stage straight through the end of a session. Such recordings may capture spontaneous ideas, dialogue and other unexpected outbursts of creativity that may add sparkle to a production. Engineers may use the monitor mix recording device or place an inexpensive, portable stereo digital recorder in the middle of the live room for such purpose.

BASIC TRACKING

Most productions are initiated with the recording of a rhythm section, a process commonly referred to as 'basic tracking'. This derives from the importance of creating a strong foundation over which more material, e.g. vocals, can be subsequently added. Standard basic tracking sessions commonly incorporate the recording of at least drums and bass, although some may include other instruments, e.g. guitars, keyboards, etc., as this may encourage artists to interact and play as if they were performing live. Recordists must pay particular attention to microphone choice and placement in circumstances where multiple instrumentalists share the same recording space, i.e. they must audition and evaluate the suitability of microphone 'leakage',

The following are a few points to be considered by the production team during tracking:

- Always try to record a reference tuning note from all instruments in isolation (when applicable).

- Avoid interrupting takes – encourage musicians to ignore small mistakes and keep playing.

- Never engage talkback during a take (unless a performer requests guidance through a difficult passage).

- Do not be afraid to 'ride' the faders in case a channel appears to be approaching the clipping point (attempt to ride the faders as fast as possible without making such changes noticeable).

- Be supportive of the performers immediately after each take is finished (consider using 'gated' talkback).

- Never let the performers detect a sense of panic in the control room (whether caused by equipment malfunction or time-related concerns).

- Never give the performers the impression that people in the control room are engaging in unrelated activities during takes, e.g. having conversations, telling jokes, etc.

- Be aware of your role and its boundaries, e.g. artists may welcome criticism from producers, but not from engineers.

- Be ready to respond to requests to improve the cue mix, although you should not let the performers make the team run in circles.

- Consider everything that the performers have to say regarding their circumstances, even if they seem to be transferring the blame for their mistakes or looking for a 'scapegoat'.

- Learn from mistakes, apologising for them in a clear and professional manner and avoid repeating them at any cost.

- Record drum 'samples' at the end of each session (in case drum replacement or extra programming is needed).

Refer to the pre-production section of this book for some extra advice on tracking, e.g. record bass amplifiers with an accompanying DI signal, record MIDI data alongside keyboard audio signals, etc.

Recording Direct to Two-Track
Certain conditions may lead engineers to opt to record directly to a two-track machine. Such approach is not uncommon in chamber music production where the channel count is small and a mixing stage is not necessary. Engineers recording directly to two-track must ensure that the stereo image of the recording is truthful and well balanced.

Compression vs. Gain Riding
In cases where the dynamic range of the material is evidently greater

Compression vs. Gain Riding (continued)

than that of the recording medium, recordists may have to choose between 'riding' faders and using a compressor. The use of dynamic range processing is common for the recording of sources such as vocals, where changes can happen too quickly and too often to be controlled via fader movements. Engineers must still consider how they would approach fader riding in order to set a compressor attack and release functions appropriately, i.e. the compressor should ideally 'mimic' what a human would do or what a human would ideally try to do (if the required attack and release are too fast).

'Gated' Talkback

The use of a 'gated' talkback function enables instant communication between live and control rooms, allowing all members of the team to hear each other's comments when the transport of multitrack machine is not engaged. Such mechanism must be used cautiously and under no circumstances should negative or derogatory remarks (humorous or not) be made in the control room.

As a quick way to generate a gated talkback function, when using a DAW, operators may create a track containing a steady tone, e.g. 1 kHz spanning over the whole duration of a song. The output of this track can be fed to the key input of a ducker applied onto the talkback channel. This way, whenever the transport of the recorder is engaged, the steady tone will reduce the level of talkback (ensure the range is set to maximum).

OVERDUBBING

Overdubbing may be described as any recording that is made onto a hard-disk session file or analogue tape that already has material on it, i.e. when new recording content is added to an ongoing production. Overdubbing is commonly used when it is not possible or advisable for multiple musicians to record simultaneously, be that due to potential microphone 'leakage' or other reasons, e.g. musicians that prefer to record alone.

It is not unusual for ensemble or band musicians to feel somewhat 'detached' or isolated when playing separately from other group members and such circumstances may lead them to perform with an inappropriate feel. In such cases it is important for the production team to help the artists connect with the pre-recorded material through the creation of an appropriate, encouraging atmosphere.

Monitoring Modes
Some recorders offer a few different monitoring modes. The following are the two most common options found in multitrack devices:

All Input / Input Monitoring
In 'all input' or 'input monitoring' mode, the recorder will output signals present at its input (as if being 'bypassed'). This mode is to be used during the initial stages of set-up only, when engineers may need to work with multitrack return signals as fast as possible, e.g. for the creation of a cue mix if 'multitrack send' signals are not available for such purpose.

Auto Input
In 'auto input':
- If the recorder's transport is not engaged, all armed tracks will output signals.
- If transport is playing, only pre-recorded signals will return from the recorder, i.e. arming a track will have no effect.
- If the transport is set on record, armed tracks will record and output signals, while all unarmed tracks will output pre-recorded material (if any is present).

Overdubbing and Tuning
It is imperative for overdubbing musicians to tune their instruments to a stable reference note from a pre-recorded instrument, as opposed to using a tuner or pitch fork.

Overdubbing in the Control Room
Recordists should consider the possibility of placing musicians in the

Overdubbing in the Control Room (continued)

control room for overdubs, as this facilitates communication. It is important for the team to evaluate the room's atmosphere and its possible effect on performance before making such decision.

'Punching In'

The process of 'punching in' is used when small sections of an otherwise good performance must be replaced, e.g. wrong words or notes. Here the engineer plays a track and drops into 'record' immediately before the section to be replaced, pressing 'play' again after the replacement. With the advent of DAWs, punching in became somewhat less common as new tracks can easily be created and used for 'comping' purposes. It is nevertheless very easy to perform punch-ins using DAWs through the use of facilities such as 'pre' and 'post-roll' and 'auto-punch'.

Bouncing

Bouncing is the process of redirecting material from the multitrack recorder to itself. In the past bouncing was a necessity, as track count was in some cases too small to accommodate sophisticated productions. In the present, the process is more commonly used when engineers want to modify recorded signals and store them again. As an example, a small studio may have one hardware EQ unit that a producer wants to use on vocals and bass. In this case the vocals may be bounced through the equaliser, freeing the device to be used on bass. NB It is important to note that the term 'bounce' may have a different meaning, e.g. 'export', when it comes to DAWs.

A FEW WORDS ON STUDIO PSYCHOLOGY

It is not uncommon for producers to find their artists acting in an unpredictable or unreasonable manner during sessions and it is important for the former to remember that the talent may feel under a lot of pressure while tracking. The recording process has an enormous importance in the life of a performer, who may or may not get to develop an artistic career depending on the success of their studio efforts. With this in mind, individuals in the control room must be fully committed and supportive of their artists, assuring them that the success of a recording session is equally important to all members of the production team (this may help distribute some of the pressure).

The following is a brief list of what may happen to recording artists under stress:

'Red Light Fever'
- Problem: Some artists may seem incapable of performing naturally and without making mistakes when the recorder is 'rolling'.
- Possible solution: Set the multitrack to record constantly, i.e. without a stop-and-go 'take' approach, and instruct the artists to perform as if the session was a rehearsal, i.e without worrying about the control room.

Unnatural Under or Over-Confidence
- Problem: Some performers may lose their confidence or become overly arrogant while tracking.
- Possible solutions:
  - Reduce the number of people in the studio.
  - Dim the control / live room lights.
  - Play the performers some of the music they love (inspiration).
  - Create an 'atmosphere' in the live room.

Change in Planned Recording Parts
- Problem: Some musicians may change their playing parts or overplay when recording, with negative results.
- Possible solutions:
  - Reduce the number of people in the studio (the musicians may be trying to impress someone or may be feeling intimidated).
  - Discuss the lyrical content and the other elements (parts) of the song and how they connect or support each other (mention a possible theatrical role of each part).
  - Bring up the original part emphasising how an instrument will sit better in the Mix By playing the agreed part.

Lack of 'Feeling'
- Problem: Artists may play their parts in a 'sterile' or 'clinical' way when recording.
- Possible solutions:
  - Discuss the lyrical content and the other elements (parts) of the song and how they connect or support each other (mention

the theatrical role of each part).
- Play the demo.
- Create an imaginative or artistic 'atmosphere' in the studio.

Lack of Objectivity

- Problem: Some musicians may lose sight of the 'big picture' while tracking, e.g. they may want to keep chasing the 'magical' or perfect take or in the case of simultaneous tracking one artist may want to discard a good take and continue recording basic tracks because of errors in their own individual performance.
- Possible solutions:
  - Play the take and highlight its positive attributes.
  - Play the take 'muting' the instrument being questioned and discuss the possibility of overdubbing parts later.
  - Suggest a break and play the recorded take upon returning to the studio (artists may judge it more objectively at this point).
  - Work on a different song temporarily and return to the original one at a later stage (with 'fresh ears').

Lack of 'Drive'

- Problem: Artists may give up too quickly when tracking and deem sub-standard performances as adequate.
- Possible solutions:
  - Suggest a break and play the recorded take upon returning to the studio (artists may judge it more objectively at this point).
  - Play the demo highlighting its positive attributes.
  - Play the performers some of the music they love (inspiration).
  - Create an 'atmosphere' in the live room and discuss the lyrical content of the track and the intention of the composer.

Aggressive Behaviour

- Problem: Performers acting unreasonably and aggressively towards each other or towards other members of the production team.
- Possible solutions:
  Call a break and:
  - Step outside the studio (address the offending members of the team individually or in small groups – attempt to diffuse situations humorously).
  - Create a distraction, e.g. take the offending members to the lounge and get them to play a video game.
  - Offer the performers something to eat or drink (some individuals become noticeably more aggressive when they are hungry).
  - Consider recording separately / overdubbing parts.

NB Any of the aforementioned problems may be caused or made worse by the presence of cameras or unrelated individuals in the recording environment.

THE MONITOR OR REFERENCE MIX

NB The term 'rough mix' is not used here, as it seems to exclude the possibility of 'release quality' monitor mixes.

During recording sessions, engineers are commonly expected to capture a number of different sound sources using a multitrack device, while creating a monitor mix onto a separate two-track machine or onto two tracks of the same multitrack recorder. The monitor or reference mix is the stereo balanced sum of all the contents of a given song and it is what the production team will take home to listen to between sessions and during the time that separates the recording and the mixing stages of a project. One may argue that the monitor mix is to the mixdown stage what the demo is to the recording stage, i.e. a frame of reference providing the team with a sense of direction, and with this in mind it is important to note that a good monitor mix may help a recordist secure the mixing of a project, while the opposite most certainly also applies. In some extreme cases, monitor mixes may end up being used for final release, so engineers must face the task of creating them with great focus and attention to detail.

Monitor mixes are in many ways no different than those that are created in dedicated mixdown sessions. They may require the use of equalisation, panning, dynamics and effects processing, although due to the sheer volume of tasks performed by engineers during recording, the creation of such mixes demands from individuals a much greater capacity to multi-task (although priority must be given to the multitrack recording in all cases).

The different aspects of mixing that should be considered by recordists during the creation of monitor mixes are discussed below.

Confidence Monitoring

The term 'confidence monitoring' is used to describe the process of listening to the playback of a recorder during tracking, i.e. the real-time monitoring of what is in effect being recorded. In the case of monitor mixes, this would imply the monitoring of the two-track recorder output, which is commonly achieved through direct patching onto the control room monitoring section of consoles (as an 'external' input) or through the monitoring of the foldback mix

> **Confidence Monitoring (continued)**
> through dedicated audio interface channel outputs (in the case of a DAW that is used for multitracking and monitor mix recording simultaneously).

FILTERING

The filtering of non-musical content, e.g. 'rumble', commonly takes place before individual signals reach the multitrack machine, i.e. engineers frequently record microphone signals through high-pass filters, with a few obvious exceptions such as bass drums, floor-toms, etc. In some cases the filtering of spectral content may seem more appropriate when applied to the monitor (Mix B) path, allowing the material to be captured unprocessed onto the multitrack recorder while being mixed musically onto a two-track device.

As mentioned before, it is important for the production team to understand the role of the different elements of a song and their contribution to the creation of a musical 'whole'. As an example, a track may have several instruments with good sounding low-end, e.g. bass drum, 808 bass drum, synth bass, bass guitar, etc., and a recordist may not be prepared to cut any of the 'bottom' on the way to the multitrack machine. In such circumstances, it would be advisable to filter the signals on the way to the two-track (monitor mix) recorder, so as to avoid an imbalance in the spectrum, i.e. the team would probably have to choose one or a couple of elements that would carry most of the low-end energy of the track.

> **Determining Filter Cut-Off Frequency**
> The knowledge of musical instrument characteristics may help engineers determine the appropriate cut-off frequency for a given source, e.g. as a traditional acoustic guitar has E2 as its lowest open string, an engineer may set a high-pass filter to cut all content below approximately 82 Hz (remember that the cut-off frequency will be attenuated by 3 dB, so in this case it should be set lower than 82 Hz).

APPLYING FILTERS (1)

Select a track to filter

In this example we will outline a possible approach to filtering an acoustic guitar track in order to remove unecessary low end or 'rumble' from the recording.

Solo the track

It is important to audition the track in isolation for filtering purposes.

Engage the 'opposite' filter

Here we will use a low-pass filter in order to determine the cut-off frequency of the high-pass filter that is needed to remove 'rumble' (set the filter at a low frequency, e.g. 75 Hz).

APPLYING FILTERS (2)

Audition what will be cut

Listen to what will be cut when the filter is switched to its high-pass function.

Sweep the filter cut-off

Move the cut-off frequency up and down until the very lowest 'usable' or desired musical content comes through, i.e. determine what is the lowest portion of content that should not be removed.

Switch the filter type

Engage the filter that was originally needed (high-pass) to cut all content below the lowest 'musical' frequency found in the take. Let's say for example that a guitarist does not play the sixth string of the instrument in a given song. It may be that the lowest fundamental intentionally produced is A2 (110 Hz) and as an engineer hears 110 Hz clearly, it will become apparent that such frequency should not be cut. A high-pass filter with cut-off set below 110 Hz should then be used.

'Rumble' and Mixing

Engineers must always consider the negative effects of 'rumble' when mixing and remember that a 'cluttered' bottom end may keep low-frequency drivers (woofers and possibly sub-woofers) from moving in a compliant and focused way, i.e. if a driver is in constant (non-musical) motion it may have difficulty tracking more instantaneous low-frequency rich content, e.g. bass drums, etc.

LEVEL BALANCING

Level balancing is probably the single most important element in mixing (in fact, in the past mixing engineers were known as 'balance engineers'). The challenge of balancing levels lies in the understanding of the roles and hierarchy of the individual elements of a production and the necessity to combine such elements into one cohesive, believable 'picture'. In the early days of recording, such balance was achieved through the placement of musicians at different distances from a single transducer and the process of recording directly to cylinder or to disc did not call or allow for what we call mixing to take place. It was only through the advent of multitrack recording that it became possible for engineers to record individual sources in isolation and without consideration for their place in a production, i.e. all instruments could now be recorded at equal, 'robust' levels and the balance between signals could be adjusted during the creation of a monitor mix and/or at a separate mixdown stage.

It may be argued that the secret to level balancing lies in maximising the impact of the different elements of a mix while making sure that they 'sit' alongside each other without conflict. For this task, engineers may adopt one of the following two strategies:

1. Work with all signals simultaneously, trying to make them fit onto a 'big picture' from the onset.
2. Start the monitor mix with one or a few signals (usually the rhythm section) and add new elements individually or in small groups.

Although a number of seasoned professionals follow and defend the first approach confidently, others prefer to mix using the second one and it may be advisable for recordists to adopt the latter, as it may be easier to implement during multitracking, i.e. when crafting monitor mixes (due to the nature of sound checking) and it may help less experienced individuals learn the basics of signal summing.

Signal Summing

It is important for recordists to understand how the elements of a production will interact when mixed to two channels and what such interaction means in terms of dynamic range. As we have seen previously, the lowest clipping point between the different components of a chain should be taken as the system's ceiling (unless gross distortion is desired). Once this figure is established, it is the task of the engineer to decide how the individual musical sources will combine and fill the available dynamic span.

It may be helpful to point out what happens when multiple signals are sent to a common destination.

- When two identical signals are summed the level reading at the summing point will be 6 dB higher than what would correspond to a single instance of the signal, e.g. when summed, two identical 1 kHz sine waves at – 20 dBFS will combine onto a 1 kHz sine wave at – 14 dBFS.

- When two non-identical signals are summed, as long as the signals are mostly coherent in polarity, the level reading at the summing point will be higher than what each individual signal would produce separately. The magnitude of the difference will depend on the degree of coherence and on the envelope of the two signals, e.g. a bass drum and a bass guitar peaking at – 6 dBFS individually may produce a combined reading approaching 0 dBFS.

With the previous points in mind, engineers should always consider what is going to happen to a monitor mix as multiple sources are brought into the picture and this should guide their early decisions regarding levels, e.g. it would make very little sense to start a mix to two-track with a bass drum producing a reading of – 6 dB under clipping if a few other elements, such as a drum machine, a bass guitar, etc., will be playing similar rhythmical patterns at comparable levels (possibly hitting downbeats together).

Individuals required to create a monitor mix while multitracking can benefit greatly from considering instrumentation and thinking ahead in terms of headroom. By bearing the basic principles of signal summing in mind, recordists may set the monitor path or DAW

> **Signal Summing (continued)**
> output faders (if mixing in the box) at reasonable, non-arbitrary levels immediately after setting the gain structure to multitrack, i.e. after checking multitrack return signals at unity gain, recordists may set the return path or output faders at what would seem to be sensible 'mix' levels. With this approach, as all instruments are sound-checked sequentially, the first components of the mix , e.g. the rhythm section, may serve as the anchor point in relation to which all other sources must conform and all components may be combined quickly into a cohesive, believable sonic landscape.

PANNING

Mixing in stereo makes it possible for engineers to distribute the energy of a production spatially (in the horizontal plane) and achieve a clarity that is arguably more difficult to attain in mono. Through the use of panning, sound sources that would otherwise fight for space in a mix may be set so they interact positively without the need for equalisation.

Although the spatial distribution of instruments in a mix should be determined on a case-by-case basis, certain elements, e.g. main vocals, bass drum and bass guitar, are consistently assigned to a predetermined position in traditional popular music, namely the centre between speakers. This is a characteristic inherited from the vinyl cutting days, where the side panning of high-energy components would commonly equate to mechanical problems during reproduction, e.g. the stylus would skip.

As a general rule in pop music, engineers should try to generate a balance between the two channels of a mix. This will ensure that listeners will not feel that they are being 'pulled' towards one of the two sides of the production, which can be distracting. In order to achieve this, one may simply think of sound sources as working in pairs and try to mirror them within the stereo spectrum. As an example, if an electric guitar is panned left, another instrument, e.g. an electric piano can be placed on the right to counterbalance the energy of the former.

If we continue exploring the approach described in the preceding level balancing section, as recordists sound check, not only can they set 'multitrack return' levels to a 'mix position', but they can also start distributing the energy of the Mix Between the left and the right of the monitor mix. Of course such distribution of elements can (and should) be re-evaluated as sessions progress, yet having an initial even image can help the production team assess the different elements of a production and verify that they are fulfilling their role in the mix.

Levels and Panning During Sound-Check

The following steps describe how a recordist may work with multitrack return signals during sound-check:

- Audition an element in isolation (channel or 'to multitrack')
- Set the correct gain structure to the multitrack recorder
- Audition the multitrack return signal at unity gain
- Set the multitrack return faders at a reasonable position (thinking in terms of the headroom of the two-track recorder)
- Pan the signal to an 'instinctive' position in the mix
- Think of another ideal element to be panned symmetrically, i.e. opposite (make a note), e.g. if the original element was a hi-hat, consider panning a shaker or a rhythm guitar opposite to it.

Panning Laws

When signals are routed to a stereo destination and panned to the centre, they will be sent with equal energy to both left and right outputs or channels. This should normally equate to a boost in overall level, as we would effectively be monitoring two versions of the same signal.

Panning laws were devised in order to avoid undesired or excessive alterations in level resulting from changes in panning. Such laws are not universal and they may attenuate centre-panned signals by 6, 4.5, 3, etc. dB (or side channels boosted by the same amount). This results from different views regarding the impact of side refections on overall levels, i.e. some theorists believe that side reflections always affect side-panned signals, so centre components should not be attenuated by the full amount of decibels that would seem intuitive.

Most DAWs offer the user the possibility of choosing the panning law to be used for a session.

EQUALISATION

Equalisation is commonly used to help improve clarity and balance in a mix. As we have seen, level balancing and panning can, and should, be used firstly for such purpose, although in many cases they may not be sufficient to generate the ideal equilibrium for a production.

The distribution of sound energy within the working frequency spectrum (broadly speaking between 20 Hz and 20 kHz) may be described as a 'vertical' process standing counterpart to the horizontal spread of panning. Such view may help engineers understand that the process of distributing energy within a mix is multidimensional and, as a result, all elements should be analysed accordingly. As an analogy, one may think of the sound generated by an instrument as a number of boxes stacked vertically with contents that vary in weight or volume. The number of such boxes is determined by the number of the instrument's distinct bands, e.g. it may seem appropriate to split an instrument into three (lows, mids and highs) or four (lows, low-mids, high-mids and highs) bands. During mixing, the contents of all the different instruments' 'boxes' must be transplanted onto a single set of shared 'storage drawers' (the mix), which must be filled evenly and with no overflow. Once again, engineers may attempt to create such balance working with a few elements in isolation, although at this point the interaction between the components of a mix must be considered from the onset.

It is important to note that some recordists will use equalisation as a means to improve the sound of a given source. In such cases, the use of different backline, microphones and techniques (including playing) will invariably yield better results and should be attempted first.

Soloing Equalised Elements

Engineers may need to employ 'unusual' equalisation in order to get an instrument to 'sit' in the mix when working on busier sounding productions. This is due to the fact that a spectrally isolated element may cut more easily through the mix (even at lower levels). It is important in such circumstances that the aforementioned instrument is not auditioned or evaluated in isolation (via solo), as it will not sound 'truthful' to the source, i.e. engineers should discourage clients to solo elements that were shaped drastically to fit a production.

The following page contains a visual example of the EQ 'boxes' analogy.

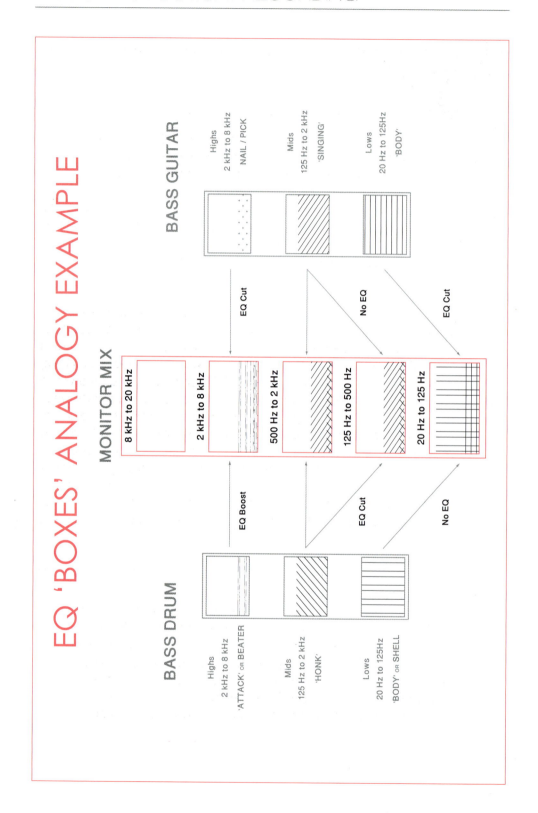

EQ 'BOXES' ANALOGY EXAMPLE

BASS GUITAR

Highs
2 kHz to 8 kHz
NAIL / PICK

Mids
125 Hz to 2 kHz
'SINGING'

Lows
20 Hz to 125Hz
'BODY'

EQ Cut

No EQ

EQ Cut

MONITOR MIX

8 kHz to 20 kHz

2 kHz to 8 kHz

500 Hz to 2 kHz

125 Hz to 500 Hz

20 Hz to 125 Hz

EQ Boost

EQ Cut

No EQ

BASS DRUM

Highs
2 kHz to 8 kHz
'ATTACK' OR BEATER

Mids
125 Hz to 2 kHz
'HONK'

Lows
20 Hz to 125Hz
'BODY' OR SHELL

NB All the cuts and boosts in the previous example are localised, i.e. they do not span over the entire band.

DYNAMIC RANGE PROCESSING

There are two main reasons that justify the use of dynamics processing:

1. The necessity to control the dynamic range of a signal
2. The desire to shape the sound of an instrument or source.

It is important to note that the first purpose is based on a need, while the second one is not. As a result, engineers reaching for a compressor, limiter, expander or gate must first establish if their use is truly necessary and, if not, what the expected outcomes of their employment are.

Compressors / Limiters

In order to determine whether a sound source needs compression or limiting due to its dynamic range, one may simply audition it throughout a production and determine whether its place in the mix seems stable or if the source seems to fluctuate excessively in level, moving in and out of the picture (this test may be particularly effective when performed at low monitoring levels). If level changes are detected, but they occur sparsely, e.g. between different sections of a song, gentle fader riding or RMS-based compression may be sufficient to control undesirable fluctuations. If level changes are pronounced and fast occurring, a combination of fader riding (automation) and peak-based compression may be required and in radical circumstances, e.g. when an instrument such as a bass guitar is expected to 'anchor' a Mix By not moving at all, limiting may be more suitable to control dynamic range.

Crest Factor and Compression

If the difference between the peak and the RMS average of a signal (the crest factor) is small, the use of compression will most likely bring no benefits (unless extreme processing is desired). If alternatively the crest factor of a signal is significant (over 20 dB), the material may benefit from the use of dynamic control. This may facilitate the mixing process in pop music as different elements are easier to place in the mix and will not shift exaggeratedly through a production.

A Musical Approach to Setting Compression Parameters

Regardless of whether a compressor is being used to control dynamic range or to shape signals, it is advisable for whoever is using such a device to examine the different envelope stages of the material that will be processed, i.e. its attack, decay, sustain and release. As a suggested approach to setting compression parameters, one could start by trying to determine whether a sound source would benefit from having more or less attack and/or more or less sustain, i.e. whether if forced to choose a technician would prefer a punchier, duller, thicker or thinner sound. Based on such observations, one may approach compression controls in a logical way and avoid the arbitrary use of dynamic range processing.

When compressors are used for the control dynamic range, they should produce undetectable results. This principle is commonly misunderstood by some novices, who feel that compressors are not doing anything unless their effects are clearly audible.

The following pages describe a suggested procedure that may help recordists use compression in a 'musical' way. It is important to point out that this procedure is more relevant and clearer when analogue devices are used (due to their less extreme attack, release and ratio settings) and when percussive sound sources are selected for processing, e.g. bass drum or snare.

USING COMPRESSORS (1)

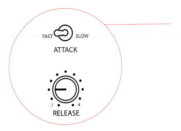

Attack and Release

Set the compressor's Attack and Release at a 'fast' setting.

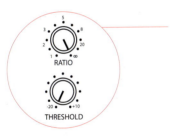

Ratio and Threshold

Set Ratio to maximum and Threshold to minimum. Audition the signal with less energy at the attack portion of the envelope (a 'choked' version of the signal).

Slow Down the Attack

Switch the Attack time to a slower setting (e.g. 100 ms). Audition the signal with more energy at the initial portion of the envelope (a 'punchier' or 'clickier' version of the signal).

Set the Attack

Choose the most appropriate Attack time (decide whether the initial stage of the original signal's envelope would benefit more from 'taming', i.e. fast attack, or from emphasis, i.e. slow attack).

USING COMPRESSORS (2)

Alter the Release

Change the Release time and audition the signal with less sustain (slow Release) or with more sustain (fast Release)

Set the Release

Choose the most appropriate Release time (fast or slow / thicker or thinner sound) – consider setting the Release time according to the interval between hits for percussive sources or choose an 'Auto' setting (if available).

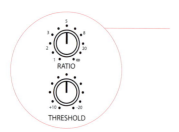

Set Ratio / Threshold

The Ratio and Threshold controls have a different influence on compression, e.g. Ratio commonly has less impact on the balance between the individual stages of the signal's envelope, while Threshold affects the signal's attack more noticeably, i.e. the attack portion ('A') of the envelope (ADSR) will be altered further. Consider changing the Threshold first when signals need further 'shaping' and use Ratio subsequently to acheive the desired amount of gain reduction.

Parallel Compression

Signals may be processed in series, where the original material is altered, or in parallel where a new processed version is created and does not replace its source. Although most of the time dynamic range control seems more appropriate when applied in series (for which it was originally designed), in some circumstances engineers may find it more suitable to compress signals in parallel.

Parallel compression allows for low level components of a track to be brought up with less impact on the overall attributes of the signal, i.e. preserving the material's transient content. This technique is commonly used to add 'size' and energy to drum recordings, where overhead or mono room microphone channels are copied and compressed at fairly extreme settings.

PARALLEL COMPRESSION (ANALOGUE)

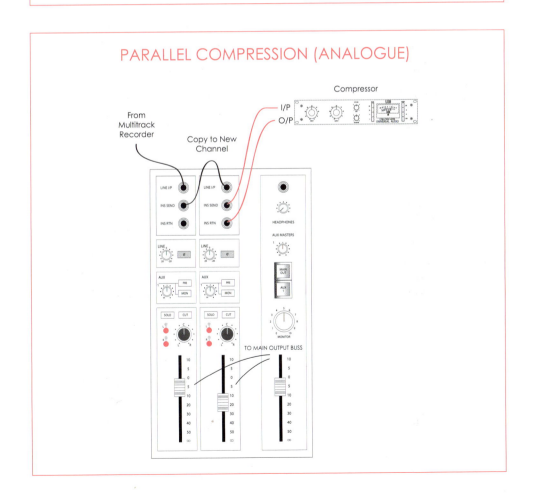

DAW PARALLEL COMPRESSION (1)

Select a Track

Choose a track to process. In this example a mono drum overhead track is used.

Route the Track

Send the original track to a new one, ensuring to choose the correct 'width' (in this case mono) and 'type' (Aux Input). Choose a sensible name for the new 'aux' track, e.g. 'OH Comp'.

Set the 'Send' Level

Bring up the 'Send' level to the new auxiliary track.

DAW PARALLEL COMPRESSION (2)

Set Up the 'Insert'

Insert a compressor onto the new auxiliary track.

Adjust the Compressor

Select an appropriate setting for the compressor.

Set Levels and Routing

Establish a good level balance between the uncompressed and compressed tracks and check that they are routed to the same output buss.

Gates / Expanders

Gates and expanders are rarely employed to process signals on their way to the multitrack recorder, as such form of dynamic range control is not commonly necessary and its misuse can lead to disastrous consequences. As far as mixing is concerned, gates and expanders are still found controlling dynamic range on occasion, although since the advent of the DAW and non-linear editing, where regions may be trimmed and levels automated, they are far more frequently used for creative purposes, such as key input (side-chain) based rhythmic gating, etc. It is important to point out that in the case of monitor mixes this must be performed quickly or simultaneously to multitracking, i.e. with no time for editing, gates may still play an important roll in 'cleaning up' tracks with microphone 'leakage', e.g. snare drum, tom-toms, etc.

Using Gates to Eliminate 'Leakage'

The following is a quick guide to using gates to reduce the effects of microphone 'leakage':

Start with the gate set as follows:
- Range at the maximum setting
- Ratio at maximum setting (if using an expander/gate)
- Attack at fastest setting
- Release at fastest setting
- Threshold at maximum setting
- Hysteresys at 10 dB (if such control is offered).

Make the following adjustments:
- Decrease the threshold until the first signals start opening the gate (just the attack portion of the envelope – the signal should be 'clicking' now.
- Increase the attack time so the signal does not click anymore.
- Lower the threshold if necessary (if the initial portion of the signal's envelope is being cut).
- Increase hold time until most of the signal envelope is going through the gate.
- Adjust the release setting so the signal decays before the first following 'undesired' signal goes through the gate.

CREATIVE GATING

Sub-Bass Oscillator

A low-frequency oscillator is used to add bottom end to a track. The sub-bass will only be heard when the bass drum is hit.

Gated Synth

Motionless synthesiser parts may be set to play rhythmically as the gate responds (opens) when the drums are hit.

The same principle may be applied to 'tighten' bass guitars that play out of time in relation to the bass drum.

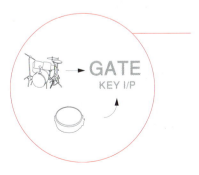

Gated Snare Reverb

The signal from the 'room' drum kit microphone is only heard when the snare drum is hit. This may help add 'size' to thinner drums.

'Duckers'

A 'ducker' may be described as a gate with an inverted function, i.e. a device that attenuates signals when an input overshoots the threshold.   Such devices are commonly used in broadcasting, where an announcer's voice forces the background music's level down.

In the absence of traditional 'duckers' (as gates that offer such function are becoming rare), engineers may use downward compressors for the same purpose, although it is important to note that the latter cannot apply a fixed amount of attenuation to signals in the same way that 'duckers' can (range).

Using Duckers with a Key Input Signal
Ensure to patch the key input appropriately and to engage the equivalent external key input function.

Start with the ducker set as follows:
- Range at the maximum setting
- Ratio at the maximum setting
- Attack at the fastest setting
- Release at the fastest setting
- Threshold at the maximum setting.

Make the following adjustments:
- Decrease the threshold level until the signal to be ducked 'disappears' when the key input signal is present.
- The signal should be clicking/pumping now.
- Increase the attack time so the signal does not click any more i.e. level reduction is applied in a musical fashion.
- Increase hold time so the signal to be ducked does not come up in level as the key input signal fluctuates briefly in amplitude, e.g. between voice-over words.
- Decrease the range so the 'ducked' signal does not become completely inaudible.
- Adjust the release setting so the signal returns to unit gain appropriately after the necessary ducking is achieved.

The following pages contain examples of creative uses for gating and ducking.

CREATIVE 'DUCKING'

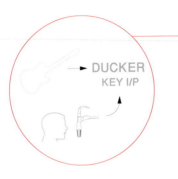

Vocals and Guitar

Whenever vocals are present the level of the guitar is attenuated. This can help in the case of songs where the vocals seem to fight for space.

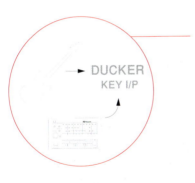

Sub-Bass Drum and Bass

Whenever a very low-frequency bass drum plays the level of the bass guitar is attenuated.
This can be particularly effective in songs that incorporate a TR808-style sub bass drum on the down-beat of the first of a group of bars.

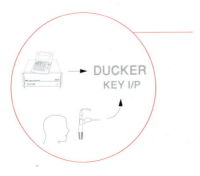

Vocals and Effects

Whenever the vocals get louder the level of their associated effects gets lower.
This can help de-clutter songs where delay or reverberation affect the clarity of the 'dry' signal.

Processing Order

Some mixing consoles make it possible for the user to source the insert send signal from a pre or post-EQ position. This can be particularly helpful when compression is to be applied in series with equalisation, as in the examples below:

- EQ Followed by Compression
  This is the ideal set-up when the dynamic range of material must be controlled within a strict range. Post-equaliser compression ensures that possible EQ boosts will not undermine the control over dynamics.

  This order of serial processing may also be explored artistically where the user purposefully overdrives the compressor (through a boost of an 'aggressive' range).

- Compression Followed by EQ
  Compression can alter the spectrum of low-end rich sound sources significantly and in the case of bass guitars and bass drums this may lead to signals that are dynamically controlled at the expense of size and tone. In such circumstances, the use of equalisation post-compression may help restore the 'size' of bass-heavy signals.

- EQ Followed by Gating
  The use of equalisation may help desensitise a gate to portions or elements of a signal, e.g. a high pass filter may be applied to a snare drum channel so that bass drum leakage does not cause the gate to open.

- Gating Followed by EQ
  Some signals require gating for the removal of 'leakage', while also benefiting from the use of equalisation. In some circumstances, the use of EQ may make the gating process difficult to achieve effectively, so dynamic range processing should ideally precede the use of EQ. As an example, a snare drum track may sound 'dull' and contain hi-hat 'leakage'. If this is the case technicians should gate the signal first and apply EQ subsequently.

EFFECTS PROCESSING

The use of effects processing during the creation of monitor mixes is largely seen as a matter of creativity, although a few important points should be considered:

- Artists will most likely expect to hear similar effects to those they heard through their headphones during recording when they audition the monitor mix (but not necessarily with the same intensity or at the same level).

- Time-based effects can add depth to a mix, making the latter more three-dimensional.

- Time-based effects may generate a sense of shared performance space to productions that rely on extensive overdubbing.

- The use of short delays may help 'thicken' sounds more efficiently than equalisation would.

Time-Based Effects and Mix Depth
The bussing of signals to a common effects processor may help engineers add a real performance 'feel' and depth to productions based on overdubs and/or recorded in very 'dry' environments. This can work particularly well with instruments that complement each other, e.g. snare drums and percussive rhythm guitars, when these are sent to the same reverb unit. Engineers should ideally use a stereo auxiliary buss to feed the effects processor, ensuring to pan both the feed to the main buss and to the 'aux' consistently, e.g. if the 'dry' signal of an instrument is panned to the left of the main mix it should ideally also be (somewhat) panned to the left of the reverb unit's input, making the overall image of the instrument in the mix credible or realistic.

Haas Effect-Based Delays
In the 1950s Helmut Haas established that if multiple versions of a given sound are offset by a small time delay, the first wavefront to arrive at the listener's position will, in most circumstances, determine

Haas Effect-Based Delays (continued)

localisation. This phenomenon is known as the Precedent Effect, Law of the First Wavefront or Haas Effect and it may be explored in mixing so as to increase the dimension of sounds. The use of short (20 to 30 millisecond) delays may be considerably more efficient in making signals appear more robust than the use of equalisation.

'HAAS EFFECT' DELAY

Increasing Instrument 'Size'

A 'thin' guitar signal may be made 'bigger' through the use of a delay effect. The delay time should be set to approximately 20 ms and the 'dry' and 'wet' signals panned opposite to each other. This technique may be used to help position a guitar that is playing alone under vocals, where the panning of the 'dry' instrument alone could make the mix left or right channel heavy.

NB The 'dry' signal should still be set at a higher level than the 'wet' one.

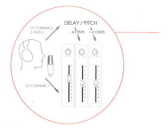

'Thickening' Vocals

Main vocals may be 'thickened' through the use of a short (Haas effect) delay and pitch shifting effects. Here the vocals are being triplicated with the right and left versions detuned by +4 and −4 cents respectively and panned opposite to each other.

> **ADT**
>
> In the 1960s, Abbey Road studios' innovative engineer Ken Scott devised a way to double up vocal tracks without the need for the performers to record multiple takes of the same part. The process involved two tape machines with unevenly spaced 'record' ('sync') and 'repro' heads, running at different speeds.
>
> The effect known as 'artificial double tracking' or ADT may be approximated through the use of digital delays (set at a time of around 40 ms) and the effect is particularly convincing if the 'dry' and 'wet' versions of the signal are panned separately and the delay times fluctuate very slightly and randomly (emulating the varispeed of tape machine transport).

SESSION INFORMATION

The recording sessions for a given project, e.g. an album, may span over days, weeks or even months. In some cases, a production team may switch between songs before their completion, which will require someone to keep track of what file, group of DAW regions or tape reel(s) to return to. Alongside this, equipment settings may be altered multiple times through the course of sessions, presenting engineers with the challenge of recalling gear parameters in order to carry on working on a given track. In order to allow for continuity during production, session notes, including recall and track sheets, etc. should be generated, allowing the team to work efficiently, saving time and avoiding costly mistakes.

RECALL SHEETS

Recall sheets are created during recording sessions so engineers can replicate certain conditions at later dates. Such documentation can be very simple, e.g. the microphone model and preamplifier gain in the case of a single transducer connected directly to an audio interface with a built-in 'mic pre', or more complex by including console and outboard gear parameters, etc. In cases where notes are very simple, the 'comments' window of a DAW application may be sufficient to store all the necessary information, with the extra benefit of no possible separation between notes and recorded material (not uncommon in the case of tape boxes and separate printed notes). Notes of greater complexity may alternatively be written on paper or be generated through the use of a computer, in the form of a text document or a dedicated software file, e.g. 'Teaboy', etc.

TRACK SHEETS AND SESSION NOTES

As a number of different recording 'takes' of a given song are completed or aborted it may become progressively more difficult for the production team to keep track of what has been committed to tape or to hard disk. It is therefore advisable for engineers to develop and use a system that describes the nature of the recorded material, conveying its quality and usability. Historically, written track sheets and session notes, kept inside analogue tape boxes (with their own set of affixed label notes), were utilised for such tasks and engineers were able to save a considerable amount of time in the studio by not having to audition tracks in order to determine their content and merit. Although the value of such documentation is inarguably greater in analogue tape based productions (with no waveform display, automatic time stamping, etc.), engineers can still benefit from having a thorough set of session notes or from simply applying the same data record-keeping principles to their DAW session files, e.g. through the sensible labelling of tracks and the consistent use of 'comment' fields. This can undoubtedly help expedite the subsequent editing and mixing stages of production and create a bond between professionals working in isolation, e.g. a good set of notes may lead a mixing engineer or producer to always want to work with a particular recordist.

Take Information

Session notes should incorporate key information describing the different takes recorded onto the multitrack device. Such information should include the song's key, tempo (plus whether a 'click' was used) and any other performance-related observations that may help engineers at subsequent stages. Session notes should also indicate the completeness of takes, where the following shorthand system may be used:

- Complete + Best or (M) – A complete take chosen as the master
- Complete – A complete (but not necessarily the best) take
- I/C – An incomplete (unfinished) take that may be used for editing
- NFU (not for use) – A take considered unusable.

The following pages contain examples of studio input / output, recall and track sheets, and session notes.

Example of Input / Output Recall Sheet

STUDIO 'X'	Date: 'X/X/XX'	Artist: 'X'		Song: 'X'		
Source	Studio I/P	External Preamp	CH	CH Name	Insert	Track
BD in / D112	D1		1	BD in	EQP1 (a)	1
BD out / U47 FET	D2		2	BD out		2
SD top / C414	D3		3	SD top	EQP1 (b)	3
SD bot / Beta 57	D4		4	SD bot		4
TT / C414	D5		5	TT		5
FT / C414	D6		6	FT		6
OH hh / 4038	D7		7	OH hh		7
OH ff / 4038	D8		8	OH ff		8
Bass / (DI)	M Tie 1	205L	9	Bass DI		9
Bass / RE20	M2		10	Bass cab		10
EGTR / MD421	M3	Red 8 (1)	11	EGTR dyn		11
EGTR / U87	M4	Red 8 (2)	12	EGTR cnd		12
AGTR / R121	M5		13	AGTR r	1176 (a)	13
AGTR / U67	M6		14	AGTR c	1176 (b)	14
KEYS L / Line	M Tie 2		15	KEYS L		15
KEYS R / Line	M Tie 3		16	KEYS R		16
HRNS L / KM84	M7	Dual Mono	17	HRNS L		17
HRNS R / KM84	M8	Dual Mono	18	HRNS R		18
BVs 1 / U87	B2	1073 (b)	19	BVs 1	2264	19
BVs 2 / U87	B2	1073 (b)	20	BVs 2	2264	20
BVs 3 / U87	B2	1073 (b)	21	BVs 3	2264	21
BVs 4 / U87	B2	1073 (b)	22	BVs 4	2264	22
VOX 1 / C12	B1	1073 (a)	23	VOX 1	LA2A	23
VOX 2 / C12	B1	1073 (a)	24	VOX 2	LA2A	24

Example of Track Sheet

STUDIO 'X' TRACK SHEET

Artist:	Artist 'X'
Song:	Song 'X'
BPM / Key:	120 / C
Session Date / Time:	'X' / 'X'
Studio:	Studio 'X'
Engineer:	Engineer 'X'
Assistant(s):	Assitant 'X'
Converter:	Converter 'X'
Reference Level:	– 18 dBfs
DAW / Version:	App. 'X' / 10.1
Sample Rate / Bit Depth:	48 kHz / 24 bit
Host Computer / OS:	Comp. 'X' / 11.1
Hard Disk:	HD 'X'

Track Content:

1 BD in	2 BD out	3 SD top	4 SD bot
5 TT	6 FT	7 OH hh	8 OH ft
9 Bass DI	10 Bass cab	11 GTR dyn	12 GTR cnd
13 AGTR R	14 AGTR C	15 KEYS L	16 KEYS R
17 Horns L	18 Horns R	19 BVs 1	20 BVs 2
21 BVs 3	22 BVs 4	23 VOX 1	24 VOX 2

Notes: Check AGTR tuning / Check SD bot for polarity.

Example of Console / Outboard Recall Sheet

STUDIO 'X' RECALL SHEET

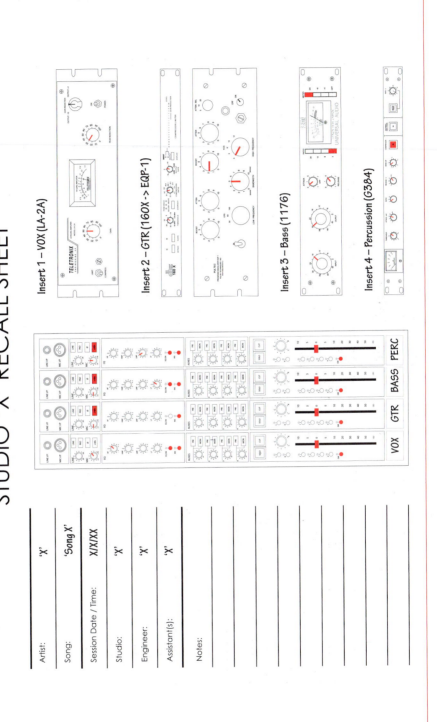

Insert 1 – VOX (LA-2A)

Insert 2 – GTR (160X -> EQP-1)

Insert 3 – Bass (1176)

Insert 4 – Percussion (G384)

VOX GTR BASS PERC

Artist: 'X'

Song: 'Song X'

Session Date / Time: X/X/XX

Studio: 'X'

Engineer: 'X'

Assistant(s):   'X'

Notes:

STUDIO 'X' SESSION NOTES

Date: X/X/XX   Producer: 'X'

Artist: 'X'   Engineer: 'X'

Label: 'X'   Assistant: 'X'

(21:00) Song 'X' – Take 1
NFU – Warm-up
Tempo: 120 BPM (no click-track)
Key: C
Form: I / V1 / C1 / INST / V2 / C2 / C3 / B1 / C4 / C5 / OUTRO
DRMS / BASS / EGTR / VOX (guide)

(21:18) Song 'X' – Take 2
I/C (first part for possible editing)
Tempo: 120 BPM (click used)
Key: C
DRMS / BASS / EGTR / VOX (guide)
Check EGTR tuning

(22:13) Song 'X' – Take 3
Complete – BEST (M)
Tempo: 120 BPM (no click)
Key: C
DRMS / BASS / EGTR / VOX (guide)
Check timing at outro

(23:35) Song 'X' – EGTR overdub
Added to Take 3
Printed with quaver delay during bridge

LYRICS SHEETS

Lyrics sheets allow for the quick communication of vocal performance-related information during tracking and overdubbing, e.g. punching-in points, doubling requirements, etc. and may also be of value during the editing and mixing stages of production. Lyrics sheets should display all sung lines of a composition clearly, i.e. they should not omit or simply indicate repeats, and they may offer detailed information regarding performance, such as breathing points, difficult passages that require a warm-up, etc. The following is an example of a lyrics sheet:

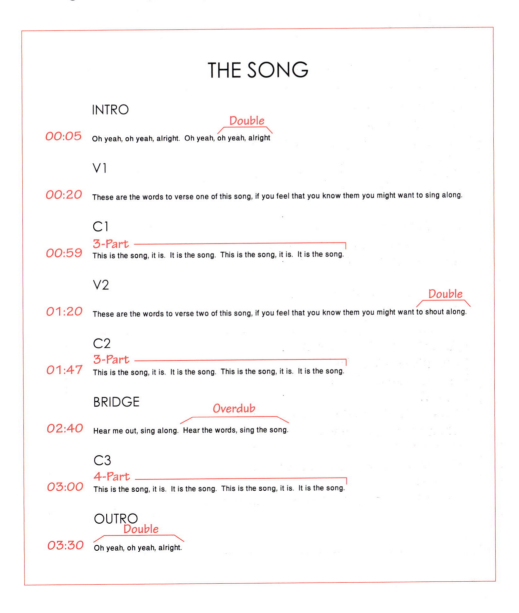

THE SONG

INTRO

Double

00:05   Oh yeah, oh yeah, alright. Oh yeah, oh yeah, alright

V1

00:20   These are the words to verse one of this song, if you feel that you know them you might want to sing along.

C1

3-Part

00:59   This is the song, it is. It is the song. This is the song, it is. It is the song.

V2

Double

01:20   These are the words to verse two of this song, if you feel that you know them you might want to shout along.

C2

3-Part

01:47   This is the song, it is. It is the song. This is the song, it is. It is the song.

BRIDGE

Overdub

02:40   Hear me out, sing along. Hear the words, sing the song.

C3

4-Part

03:00   This is the song, it is. It is the song. This is the song, it is. It is the song.

OUTRO

Double

03:30   Oh yeah, oh yeah, alright.

The Scribble Strip

It is easy for engineers to lose track of what is feeding each of the channels (and possibly the monitor paths) of a mixing console when numerous sound sources are recorded simultaneously. In such circumstances, it is advisable for recordists to create 'scribble strips', following a labelling system based on instrument name abbreviations written on high-quality (non-residual) masking tape.

Scribble Strip Abbreviations

The following are a few suggested scribble strip abbreviations:

ABASS – Acoustic bass
AGTR – Acoustic guitar
ASAX – Alto saxophone
BD in – Bass drum (inside of shell)
BD out – Bass drum (outside of shell)
BSAX – Baritone saxophone
BSSN – Bassoon
BVs – Backing vocals
CLR – Clarinet
CNGS – Congas
EBASS – Electric bass
EGTR – Electric guitar
EPNO – Electric piano
FHRN or HRN – French horn
FLT – Flute
FT – Floor tom
GPNO (b) – Grand piano (bass)
GPNO (tr) – Grand piano (treble)
HH – Hi-hat
OH ft – Drums' overhead (floor tom side)
OH hh – Drums' overhead (high-hat side)
PERC – Percussion
RTB – Return talkback
SD bot or SD b – Snare drum (bottom)
SD top or SD t – Snare drum (top)
SHKR – Shaker
SSAX – Soprano saxophone
SYNTH – Synthesiser

Scribble Strip Abbreviations (continued)
TAMB – Tambourine
TMPN – Timpani
TRMB – Trombone
. TPT – Trumpet
TSAX – Tenor saxophone
TT – Tom-Tom (possibly TIM, TAM and TOM if three toms are used).
UPNO (b) – Upright piano (bass)
UPNO (tr) – Upright piano (treble)
VCL – Violoncello
VLA – Viola
VLN – Violin
VOX – Main vocals

NB Instruments with short names such as the tuba, oboe, etc. do not require the use of an abbreviation.

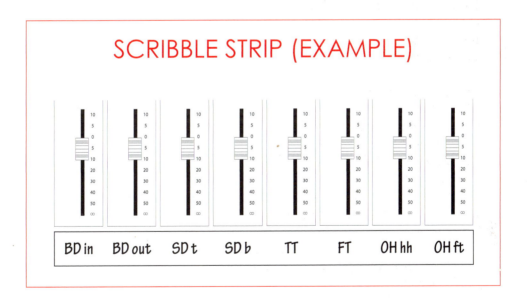

SCRIBBLE STRIP (EXAMPLE)

| BD in | BD out | SD t | SD b | TT | FT | OH hh | OH ft |

THE PRODUCTION PROGRESSION CHART

It is crucial for producers to have a clear vision of their progress through the recording stage of a project. This can be particularly challenging during sessions that span over extended lengths of time and that require multiple overdubs. In such conditions it is common for production teams to utilise some form of visual tool outlining what has been accomplished and what is still needed for the completion of a project.

The use of charts and other visual aid devices may help the team focus and work towards a given deadline from the beginning of a production (and not only when the end is imminent). The following are two possible approaches to devising a 'progression chart'. The first one provides a more global view while the second is song specific. Both approaches share the following principles:

- Empty boxes are filled as material is recorded, auditioned and approved.
- A solid grey box indicates that a particular instrument will not be used in a song.
- A letter 'G' inside a box indicates that a part was recorded, although it will only be used as a 'guide', e.g. 'scratch' vocals.
- A diagonal line or a tick inside a box indicates that a part has been recorded, although more overdubs may be necessary or the part may require further assessment.
- A solid black box indicates that all the corresponding instrument parts for a given song have been recorded successfully / been approved.

PRODUCTION CHART (1)

PROJECT NAME	DRMS	BASS	AGTR	EGTR	KEYS	PERC	VOX	BVs
SONG 1					■			
SONG 2					■			
SONG 3	■	■	■	■				
SONG 4	■		■	■				
SONG 5								
SONG 6								
SONG 7					■			
SONG 8				■				
SONG 9					■			

PRODUCTION CHART (2)

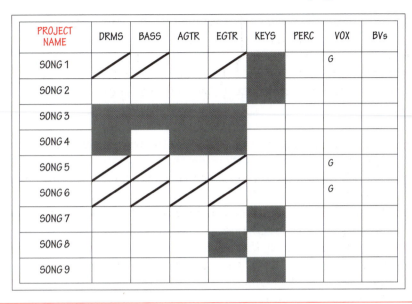

PRODUCTION CHART (3)

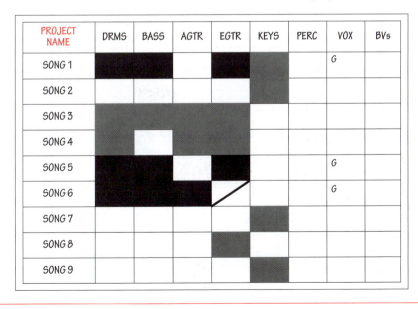

From the first chart, we can see that during pre-production the team decided that:

- Songs 1, 2, 7 and 9 would have no keyboard parts.
- Song 3 would feature keyboards, percussion and vocals (including BVs) only.
- Song 4 would be similar in instrumentation to Song 3, with an added bass part.
- Song 8 would not have an electric guitar part.

From the second chart we can gather that the during the first day of production, drums, bass, electric guitars and guide vocal parts were recorded for songs 1, 5 and 6 and an acoustic guitar part was captured for song 6.

From the third chart we can see that the team has finished and approved the drums and bass parts for songs 1, 5 and 6. The same applies to the electric guitar, although some extra parts need to be overdubbed onto song 6.

SONG-BASED PROGRESSION CHARTS

A recording team may choose to work with multiple song-based charts when tracking the progress of a production. Such charts should outline the instrumentation used in the different sections of a composition, providing the team with a more detailed view of what has been accomplished and what still needs to be done on a song-by-song basis. The following is an example of a song-based progression chart, which is being filled out as parts are recorded:

SONG-BASED PRODUCTION CHART (1)

SONG 1	INTRO	V1	V2	C1	V3	C2	BRIDGE	C3	C4	OUTRO
DRUMS	■						■			■
BASS	■						■			■
AGTR										
EGTR	■									
EPNO		■	■							
PERC	■		■							
SAX	■	■	■	■	■	■				
VOX										
BVs	■	■	■							

SONG-BASED PRODUCTION CHART (2)

SONG 1	INTRO	V1	V2	C1	V3	C2	BRIDGE	C3	C4	OUTRO
DRUMS	■	✓	✓	✓	✓	✓	■	✓	✓	■
BASS	■	✓	✓	✓	✓	✓	■	✓	✓	■
AGTR	✓	✓	✓	✓	✓	✓	✓	✓	✓	✓
EGTR	■									
EPNO	■	■	■							
PERC	■	■								
SAX	■	■	■	■	■	■	■			
VOX	G	G	G	G	G	G	G	G	G	G
BVs	■	■	■							

SONG-BASED PRODUCTION CHART (3)

SONG 1	INTRO	V1	V2	C1	V3	C2	BRIDGE	C3	C4	OUTRO
DRUMS	■	█	█	█	█	█	■	█	█	■
BASS	■	█	█	█	█	█	■	█	█	■
AGTR	█	█	█	█	█	█	█	█	█	█
EGTR	■	✓	✓	✓	✓	✓	✓	✓	✓	✓
EPNO	■	■	■	✓	✓	✓	✓	✓	✓	✓
PERC	■	■								
SAX	■	■	■	■	■	■	■			
VOX	G	G	G	G	G	G	G	G	G	G
BVs	■	■	■							

From the first chart we can gather that during pre-production for 'Song 1' the team decided that:

- The introduction should feature main vocals and acoustic guitar only.

- The drums, bass and electric guitar enter on verse 1.
- The percussion starts playing on verse 2.
- Electric piano and backing vocals parts are added from chorus 1.
- A saxophone plays from the bridge.
- The bass and the drums stop playing during the bridge and again at the outro.

From the second chart, which corresponds to the first day of recording, we can see that the drums, bass, and electric guitar parts were tracked successfully (plus a guide or 'scratch' vocal take) and from the third chart we can gather that during the second session electric guitar and piano parts were added to the production.

From the third chart we can gather that all the drum, bass and acoustic guitar parts recorded on day 1 were approved and Song 1 does not require any extra takes of the aforementioned instruments.

The team would subsequently continue to fill the chart until all part boxes were blackened out, which should mark the completion of recording for 'Song 1'.

Progression Charts and the Deadline
The use of progression charts may be of great value in the case of productions that must be completed within a very limited amount of time. With the help of such simple visual tools, a recording team can work dynamically, reacting quickly to adversities. As an example, producers and artists may agree to create a simpler, more basic version of a song if time is running out and overdubbing is not possible. Such cases emphasise the importance of having a plan 'B' and being ready to sacrifice certain ideas in order to finish a production on time and within budget.

Backup and Delivery
It is important to adhere to a consistent system for the labelling and storage of files. As a recommended practice, recordists should try to keep their complete recording DAW session files backed-up and perform edits on a copy of the master. Upon completion of the editing stage another major backup should be performed and the

Backup and Delivery (continued)
mixing sessions should ideally run from a copy of the final 'edited' DAW session file.

Hard Drives
Due to the possibility of hard-drive failure, recordists should ideally back up their session materials onto at least two devices (creating a 'main' backup and a 'safety' copy). It is important to use ('spin') such hard-drives at regular intervals as this may help to avoid mechanical problems and loss of data.

With the development of more reliable solid-state drives, the creation of a 'safety' copy may seem unnecessary, although it is important to remember that all media have a given life span and files may become corrupt with the passing of time.

END OF RECORDING

The final day of a project's recording stage is possibly the last moment when all the key players of a production are found together in one location. This date should be celebrated and memories, plans, etc. should be shared. With luck the production team will meet again, forging stronger relationships and generating new output.

MUSIC PRODUCTION

RECORDING

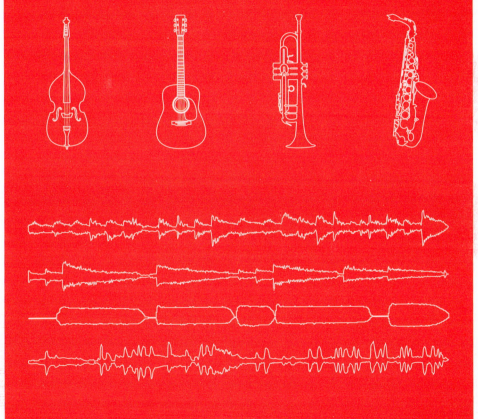

OUTRO

IS IT WORTH IT?

What makes something worth recording? With the advances in DAW technology, including the development of virtual instruments and through recent changes in manufacturing processes, it has become progressively easier for music makers to commit their ideas onto some form of medium that is ready for duplication. In addition to this, through the immediacy of Internet distribution, such ideas can be made available for public appreciation within minutes of conception.

We may say that all music has value, no matter how simple or derivative, although some may question whether all music is worth sharing. If a sound file is like a picture, what makes it worth showing to others? In the past, recording was mostly a collective effort, with teams of composers, performers, engineers, producers, etc. working towards a (mostly) common goal. This, like the case with writers, directors, editors and producers, in some instances helped widen the scope of a final product that represented the combined effort of many individuals (perhaps functioning as the diary of a 'tribe'). At the same time, throughout history, some artists have preferred to work in isolation or in very small groups, aiming to preserve the purity of the their vision. Regardless of the size of a production team, it seems that the most resilient examples of art in the form of recorded sound came from just that: a 'vision' or the need to express an emotion.

It may be said that in the past decades, due to the development and refining of non-linear editing tools and through a perceptual shift regarding the importance of the mixing stage, the general approach of recordists towards their craft has changed. It is currently not uncommon for producers, engineers and musicians to settle for 'average' performances rather quickly, as these may be 'improved' subsequently through tuning, time stretching, looping, etc. Maybe some of these individuals would benefit from listening to the multitrack recordings of songs that are regarded as 'classics', where a great mix is at times achieved when the faders are simply brought up.

To answer the question of what makes something worth recording, we should perhaps revisit the idea that to commit a song to a medium means to conserve it, to keep it from being lost. With this in mind, as long as a performance has meaning it deserves to be captured and this holds true even in cases when the performer is the only one who can see it. As a final thought, maybe the secret to 'good' music production lies in adding meaning to the recorded moment, or reminding those involved in the process that everything may in fact be unique and that the final product, much like a picture, will serve as a register of who they were at a given moment in time.

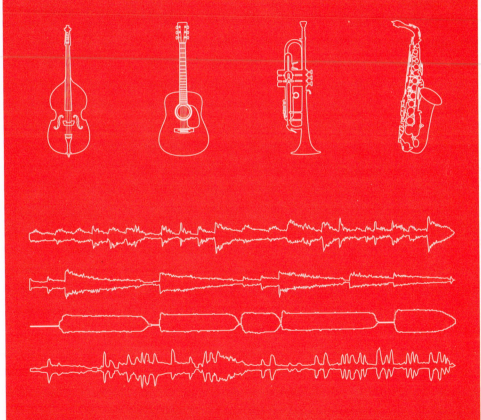

MUSIC PRODUCTION

RECORDING

APPENDIX 1
WORKING WITH REFERENCES

WORKING WITH REFERENCES

Visit a prestigious museum and there is a chance you will find art students copying pieces of artwork on display onto their sketch pads. It is not difficult for an observer to deduce the benefits of such activity, as a critical comparison to an unchanging frame of reference is a reliable way to measure the development of one's technique.

In sound engineering it is not unusual for students to underestimate the importance of referencing, when such a simple activity can do much more for an aspiring engineer than the opinion of peers, teachers, etc. The following suggested exercise is based on the assessment of 'original' vs. 'emulation' sounds and may help recordists develop their technique:

Referencing Exercise:
- Choose an instrument, e.g. the acoustic guitar.
- Choose a well-established production as a reference, e.g. Paul McCartney's *Jenny Wren*.
- Try to recreate the chosen guitar sound (not necessarily using the same song).
- Compare your recording with the reference.
- Establish possible changes in recording technique that may make the two recordings closer in sound, e.g. the use of an alternative microphone, different placement, etc.
- Apply the changes, record and reference again.

A precise match with the reference is not essential (or desired) here, as the benefits of this exercise should derive from the process itself, i.e. this activity should help recordists develop the ability to react to inadequate results through sensible changes in approach or technique.

Extensive referencing may ultimately lead engineers to devise strategies to achieve the sounds they 'hear' in their minds, an invaluable skill in music production.

The following are a few last thoughts on referencing:
- Referencing does not necessarily lead to the production of overly derivative art.
- The combination of multiple influences may lead to the creation of innovative music.
- An influence may be something one chooses to follow, or something one is aware of, choosing to break away from it.

MUSIC PRODUCTION

RECORDING

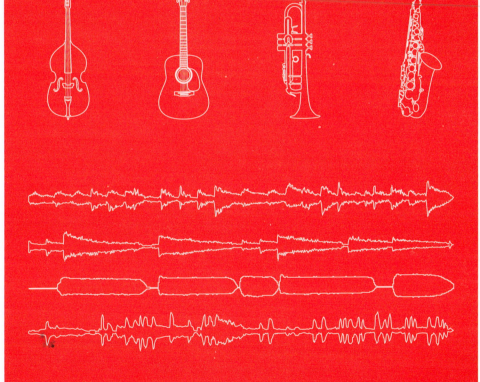

APPENDIX 2

BAR CHARTS

BAR CHARTS

Although simplified charts outlining song structure, such as the ones found in the Arrangement/Instrumentation section of this book, may be sufficient to help producers run standard pop recording sessions with some degree of control over performance, it may be useful to refine the song charts further so as to increase the degree of precision in communication.

The following examples display previously seen songs broken down into bars:

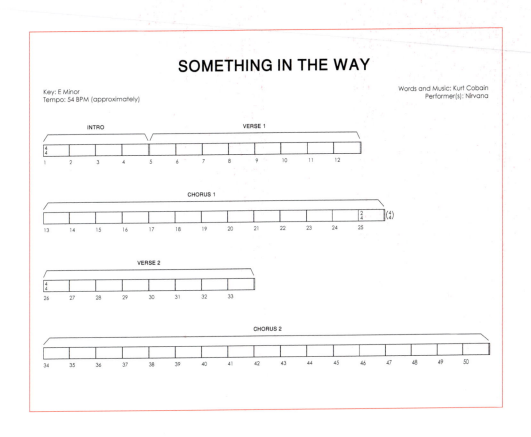

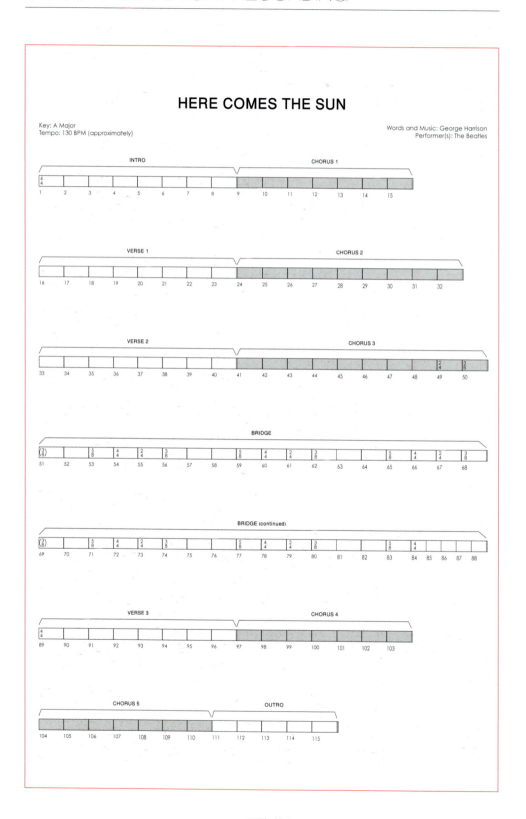

MUSIC PRODUCTION

RECORDING

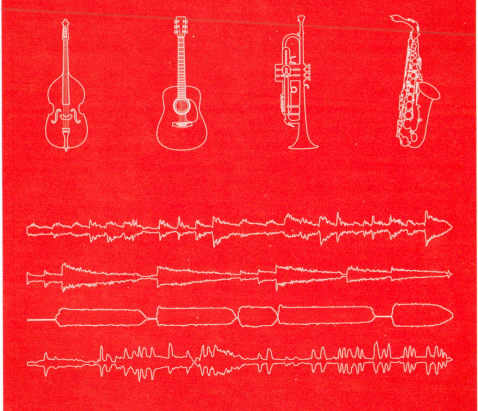

APPENDIX 3
EQUAL LOUDNESS CONTOURS

Equal Loudness Contours

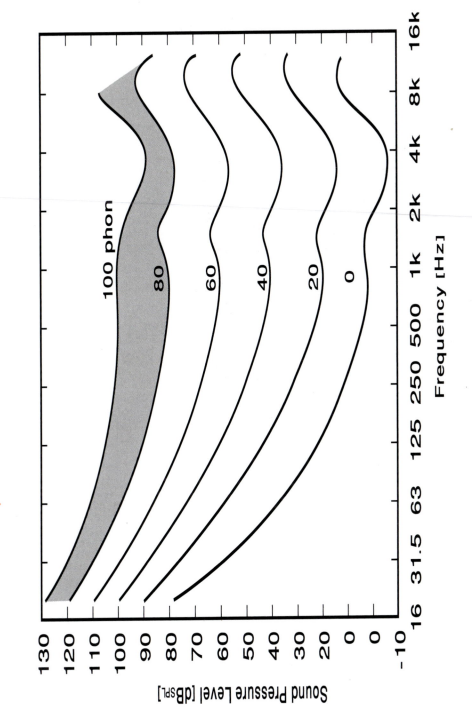

MUSIC PRODUCTION

RECORDING

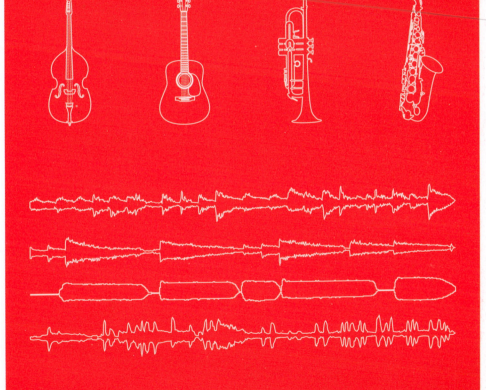

APPENDIX 4
INSTRUMENT PART NAMES

BASSOON

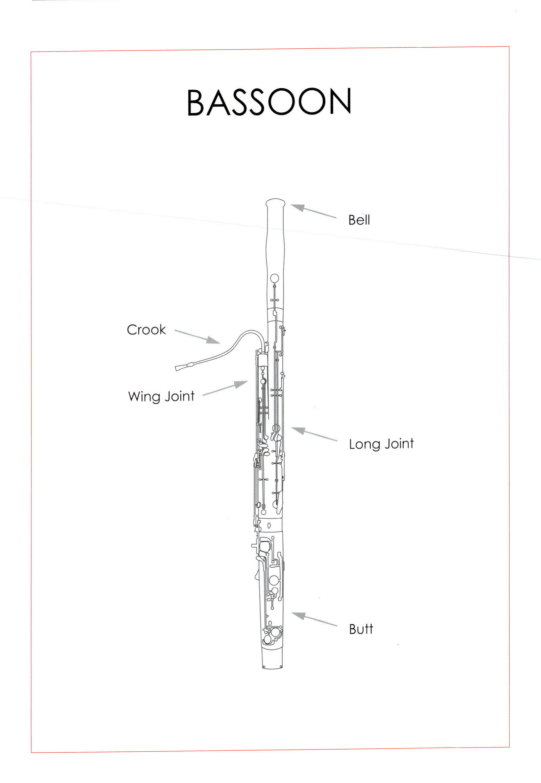

Bell

Crook

Wing Joint

Long Joint

Butt

CLARINET

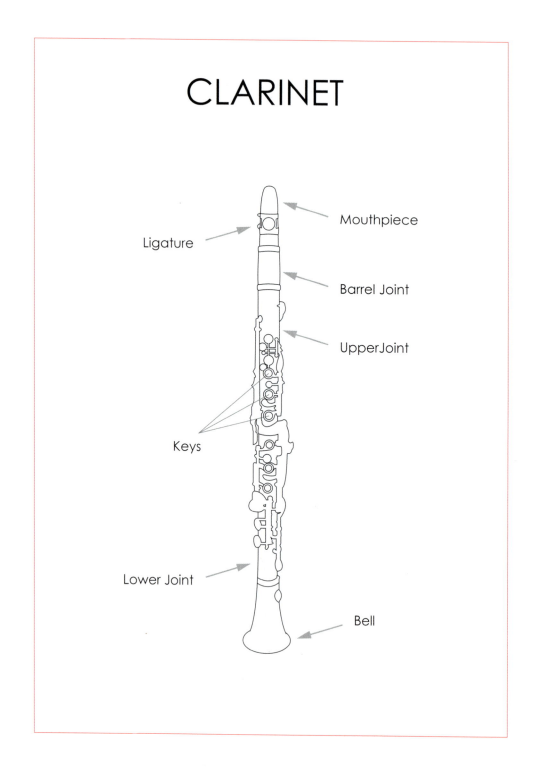

Mouthpiece

Ligature

Barrel Joint

UpperJoint

Keys

Lower Joint

Bell

FLUTE

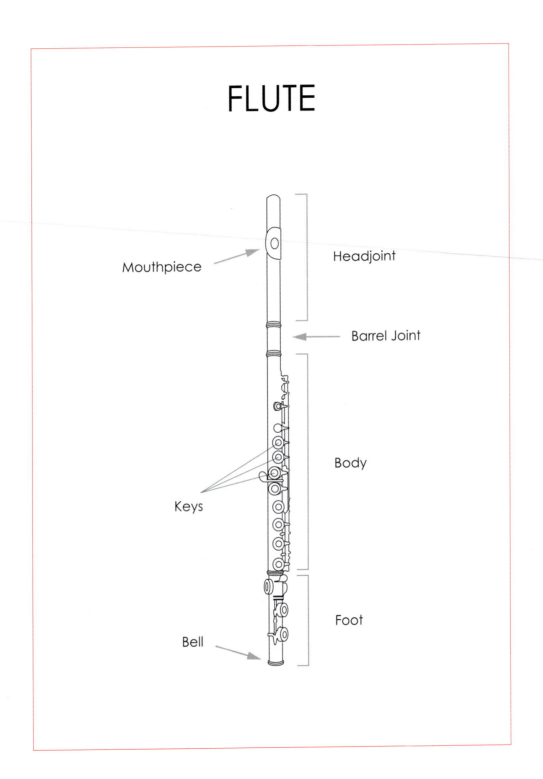

Mouthpiece

Headjoint

Barrel Joint

Keys

Body

Foot

Bell

FRENCH HORN

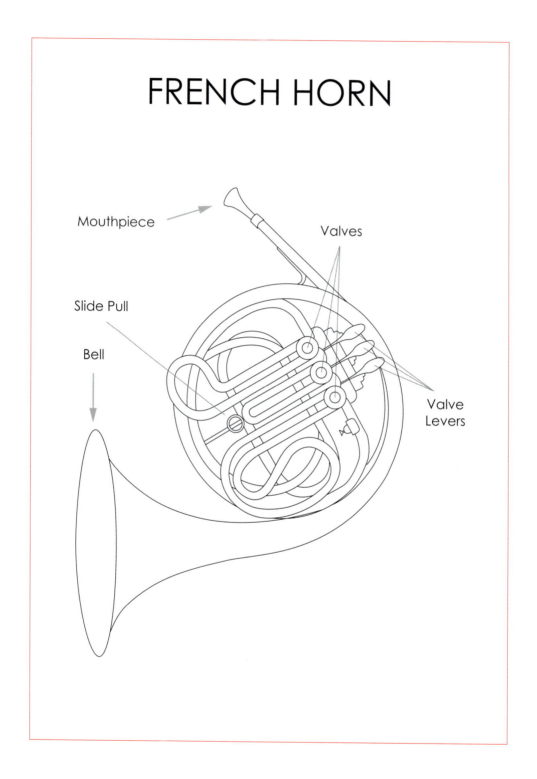

Mouthpiece

Valves

Slide Pull

Bell

Valve
Levers

GLOCKENSPIEL

Metal Bars

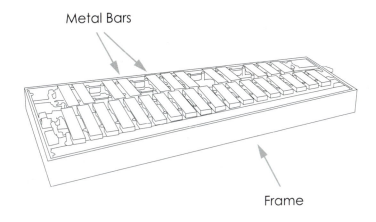

Frame

GUITAR

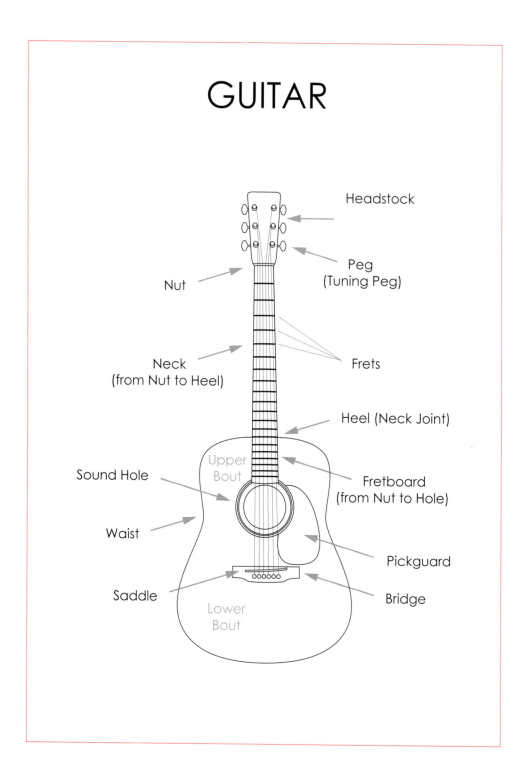

ELECTRIC BASS

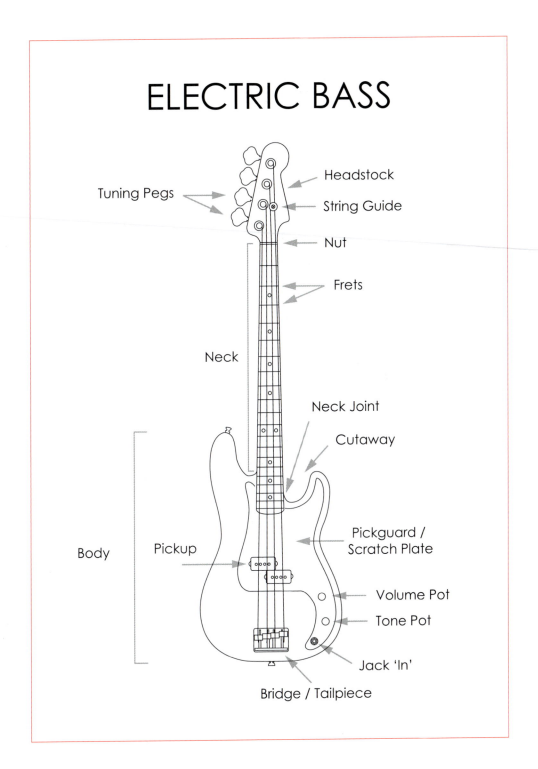

ELECTRIC GUITAR

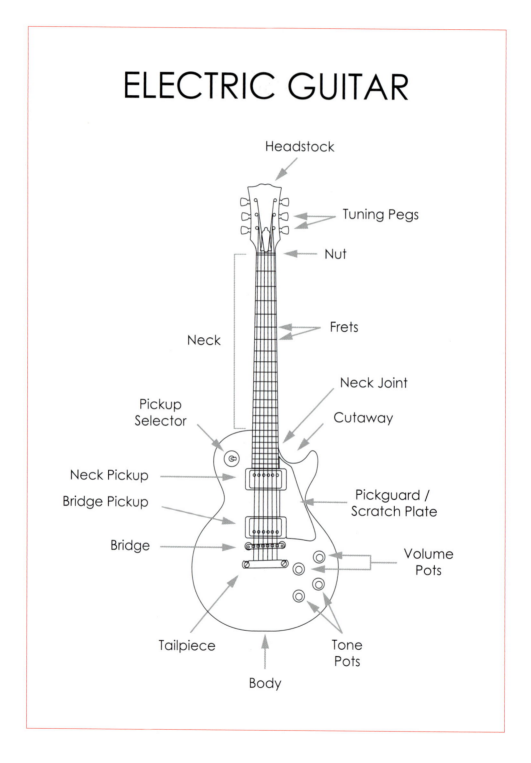

Headstock

Tuning Pegs

Nut

Frets

Neck

Neck Joint

Cutaway

Pickup Selector

Neck Pickup

Bridge Pickup

Bridge

Pickguard / Scratch Plate

Volume Pots

Tailpiece

Tone Pots

Body

OBOE

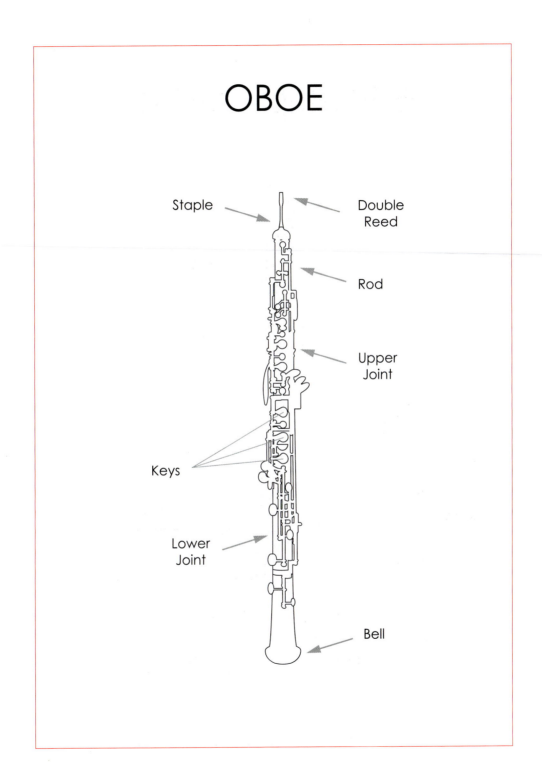

Staple

Double Reed

Rod

Upper Joint

Keys

Lower Joint

Bell

PERCUSSION BEATERS

Sticks

Rutes

Mallets

Brushes

Dreadlocks

PIANO

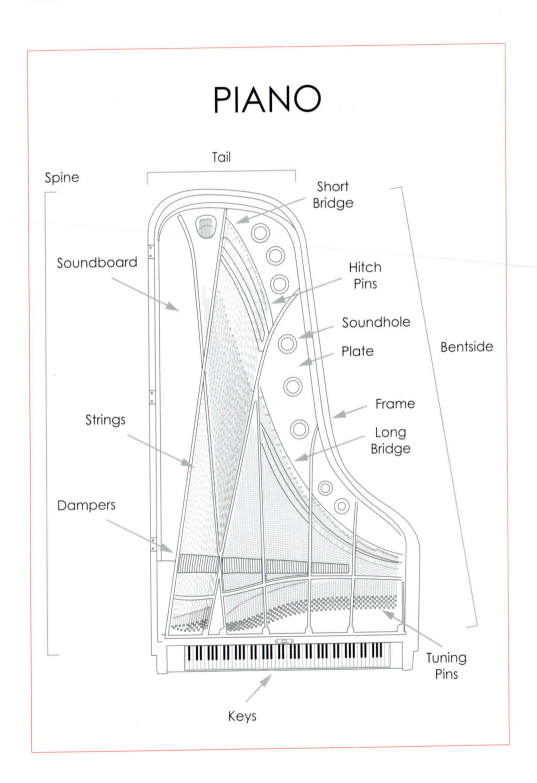

Tail

Spine

Short Bridge

Soundboard

Hitch Pins

Soundhole

Plate

Bentside

Strings

Frame

Long Bridge

Dampers

Tuning Pins

Keys

SAXOPHONE

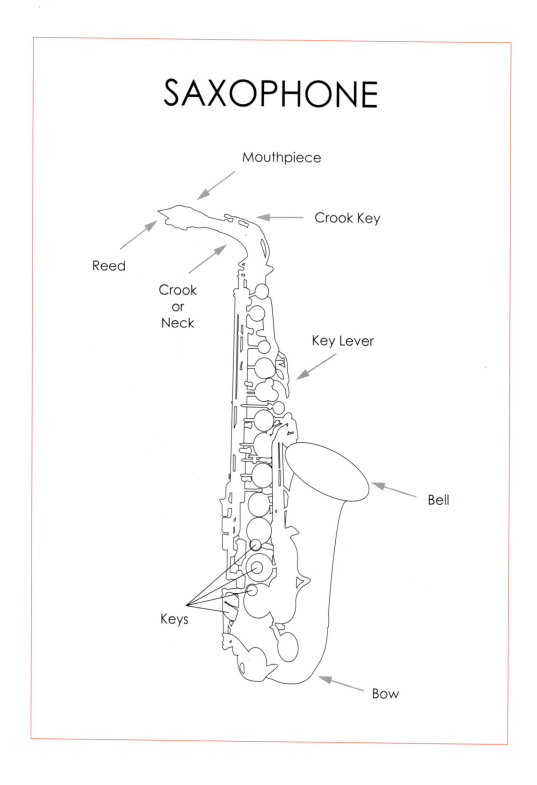

Mouthpiece

Crook Key

Reed

Crook
or
Neck

Key Lever

Bell

Keys

Bow

SNARE DRUM

(upside down)

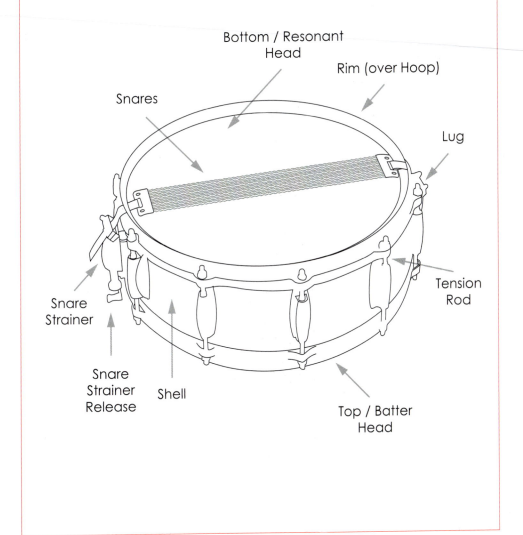

Bottom / Resonant Head

Rim (over Hoop)

Snares

Lug

Snare Strainer

Snare Strainer Release

Shell

Tension Rod

Top / Batter Head

TOM-TOM

Top / Batter
Head

Rim (over Hoop)

Tension
Rod

Lug

Shell

Bottom / Resonant
Head

TROMBONE

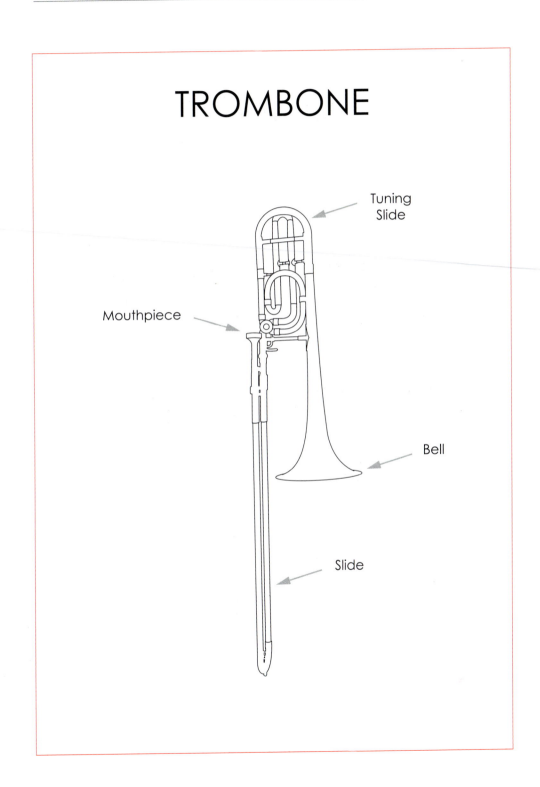

Tuning
Slide

Mouthpiece

Bell

Slide

TRUMPET

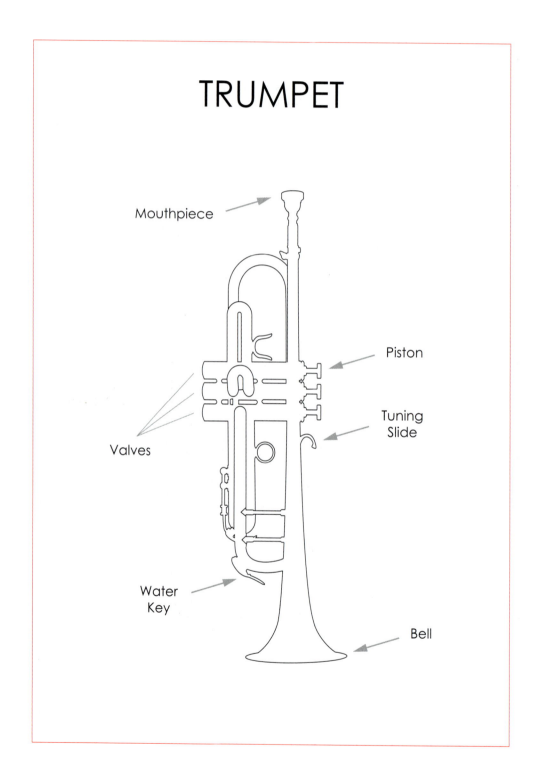

Mouthpiece

Piston

Tuning
Slide

Valves

Water
Key

Bell

TUBA

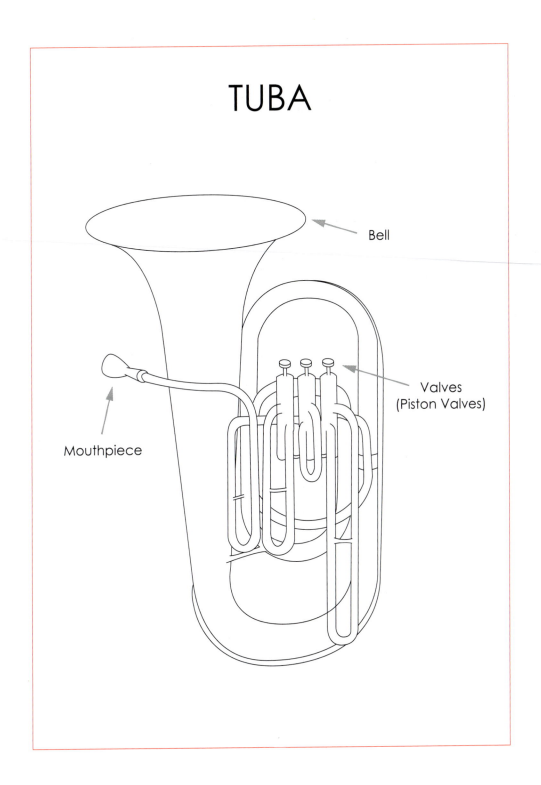

Bell

Valves
(Piston Valves)

Mouthpiece

VIBRAPHONE

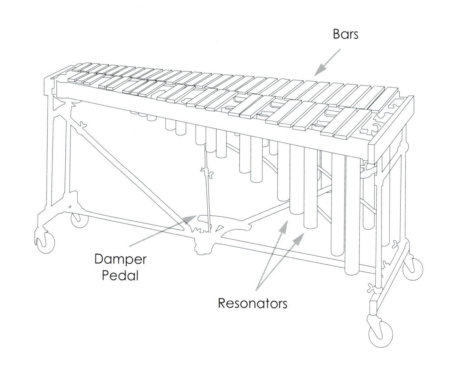

Bars

Damper
Pedal

Resonators

VIOLIN / VIOLA
CELLO / BASS

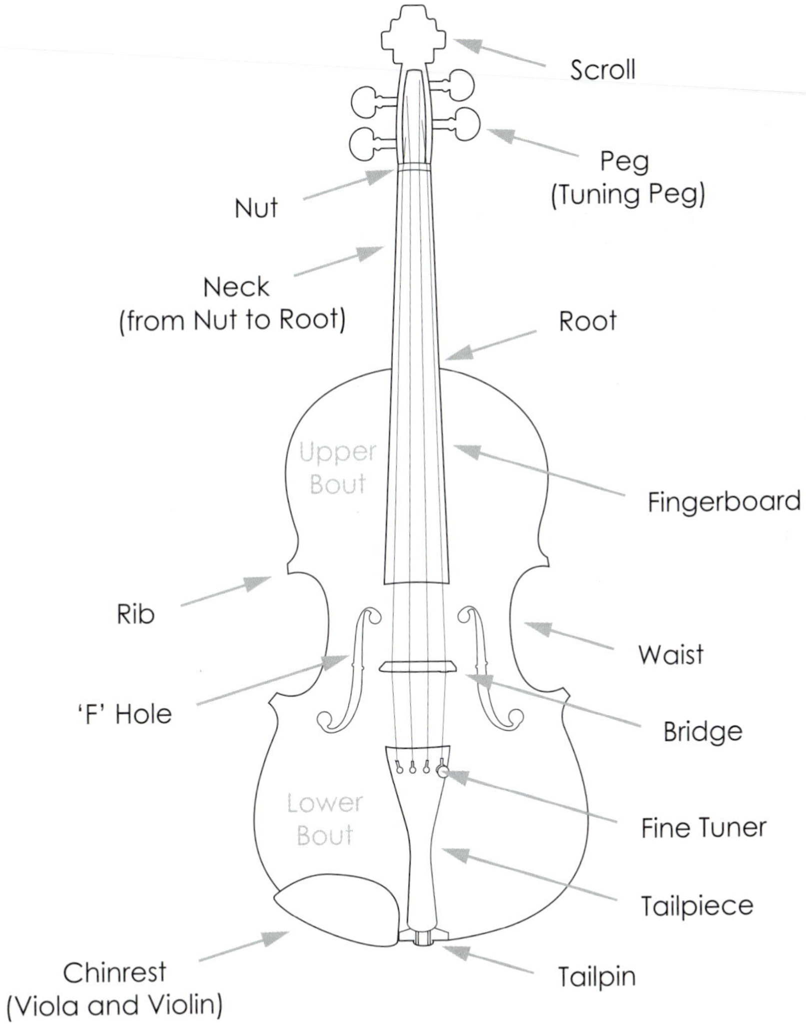

Scroll

Peg
(Tuning Peg)

Nut

Neck
(from Nut to Root)

Root

Upper
Bout

Fingerboard

Rib

Waist

'F' Hole

Bridge

Fine Tuner

Lower
Bout

Tailpiece

Chinrest
(Viola and Violin)

Tailpin

MUSIC PRODUCTION

RECORDING

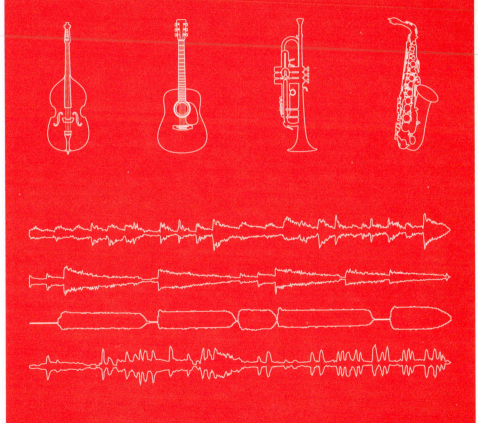

APPENDIX 5

SETTING GAIN STRUCTURE AND ROUTING SIGNALS

SETTING GAIN STRUCTURE AND ROUTING SIGNALS

The following is a description of the steps necessary for the recording of a single microphone signal through the use of a typical large-format, analogue in-line console:

- Ensure the control room output is 'muted'.
- Set the console onto 'Record' status (if such function is offered).
- Assign the preferred faders to the channel and the monitor paths, i.e. choose whether to engage 'Fader Reverse'.
- Set the threshold for the channel 'Signal Presence' light (LED) to its minimum level.
- Set the module meters to display 'Channel Input' levels, e.g. 'Channel to Meters' or 'Channel I/P' (pre-fader channel metering).
- Check the live room patching information, i.e. the number of the patch point being used.
- Set the corresponding channel input to 'Microphone' or 'Mic' (possibly through an input 'Flip' switch).
- Engage Phantom Power if necessary.
- Ask an assistant to speak onto the microphone continuously.
- Check for signal presence (LED).
- Apply gain to the microphone preamplifier and check the module meters (do not worry about levels too much at this point, the assistant is not the artist).
- Ensure the module's equaliser and dynamics processor are not engaged.
- Ensure the 'Insert In' function is not engaged (check that there is nothing patched into 'Insert Return').
- Consider performing a 'PFL' solo at this stage to audition the signal pre-fader.
- Raise the channel path fader to unity gain (marked by the letter 'U' or the number zero).
- Assign the signal to a group buss (alternatively choose to feed the recorder via 'Direct' function).
- Check the panning at the routing matrix (disengage panning if not in use).
- Ensure the group buss fader is at unity gain.
- Meter the group buss or 'direct' output signal, i.e. set the I/O module meters to 'Off-Buss' or 'DAW Send' or 'Send').
- Arm the corresponding recorder track.
- Meter the recorder's input signal.

- Set the module meters to 'Off-Tape' or 'DAW Return'.
- Raise the monitor path fader to unity gain (marked by the letter 'U' or the number zero).
- Assign the monitor path signal to the 'Main Output' or 'Stereo Buss'.
- Set panning to the desired position.
- Raise the 'Main Output' buss fader to unity gain.
- Meter the 'Main Output' buss signal.
- Source the 'Main Output' buss signal at the Control Room Monitoring section.
- Select the speakers to be used.
- Disengage the console's output 'Mute' switch.
- Ask the performer to sing or play the part he or she will be recording at the appropriate intensity (drummers should play the whole kit even if the recordists are setting up the gain structure for a single element, e.g. the bass drum).
- Check the headroom at the loudest passage (leave a few decibels below clipping free).
- Check levels at the quietest passage (determine if compression is needed).

MUSIC PRODUCTION

RECORDING

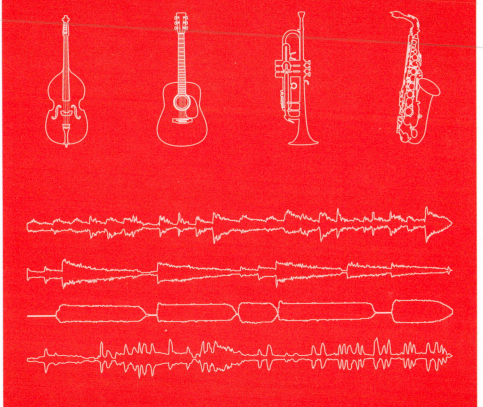

APPENDIX 6
DRUM TUNING AND TIMBRE

DRUM TUNING AND TIMBRE

It is not uncommon for engineers or assistants to be required to replace heads and tune drums in recording studios.

The following steps outline head replacement and drum tuning procedures:

- Remove the rim of the drum and the head to be replaced.
- Check the drum shell (particularly the hoop) and make sure it is not warped.
- Clean the hoop with a dry cloth.
- Check the sound of the new head (tap on it and check that its tone is not 'dead')
- Place the new head onto the hoop (do not force it).
- Place the rim over the new head and insert tension rods.
- Tighten the tension rods as far as possible by hand, using a 'counter-lug' sequence, e.g. 12 o'clock, 6 o'clock, 2 o'clock, 8 o'clock, 4 o'clock and 10 o'clock for a drum with six lugs.
- Push down the rim around each lug and tighten the tension rods further by hand (still using a 'counter-lug' sequence).
- Push down the centre of the head slowly until you hear it 'crackle'.
- Tighten each tension rod further by a quarter to half a turn using a drum key or wrench (use the same 'counter-lug' sequence) until the head has no 'wrinkles' left.
- Hit the centre of the drum in order to check its pitch.
- Tighten each tension rod further by an eighth to a quarter turn until the desired pitch is achieved (this should be the point where the drum appears to 'sing').
- If a pitch 'beating' or undesirable overtones are produced:
  - Tap the area immediately above each lug and check for differences in pitch.
  - Add tension or loosen individual tension rods until the same pitch is produced throughout.

NB Always disengage the snares when tuning.

TUNING DRUMS WITH NO LUGS OR ROPE

Hand drums with no lugs or rope-based tuning system may be pitched up through the use of heat. An electric blanket, a blow-drier or the sun may be used for such purpose, although it is important to note that the effect will be temporary and the drum will require retuning at regular intervals.

NOTES ON DRUM TUNING / TIMBRE

Bass Drums

- Bass drums are usually tuned to their lowest stable pitch or to a neighbouring tone that is sympathetic to the key of a given song (commonly between C1 and C2).
- The timbre of a bass drum is significantly affected by damping.

Snare Drum

- The top head of the snare drum is usually tuned a third, forth or fifth above the bottom head.
- The heads are commonly tuned between 'F' and 'B', e.g. bottom at 'G' and top at 'B'.
- Overly tight bottom heads may cause the metal snares to sound loose and unfocused, although this may seem counter-intuitive.

Tom-Toms

- The two heads of tom-toms are commonly tuned to the same pitch or the batter (top) head is tuned between a third and a fifth above the resonant (bottom) head. Alternatively the top head may be tuned lower than the resonant head, if a drier sound is desired.
- Tom-toms may be tuned at regular intervals, e.g. fourths, from each other. This may be particularly effective if the top and bottom heads of each tom-tom are also offset by a fourth, e.g. the pitch of the bottom of the high rack tom matches that of the top of the middle tom-tom, etc. Alternatively tom-toms may be tuned to the most important degrees of a given song's key.

Hi-Hats

The thickness of hi-hats dictates the overall timbre of the instrument, thicker cymbals tend to sound 'darker' and louder, while thinner ones sound brighter and more focused and are easier to record and mix.

MUSIC PRODUCTION
RECORDING

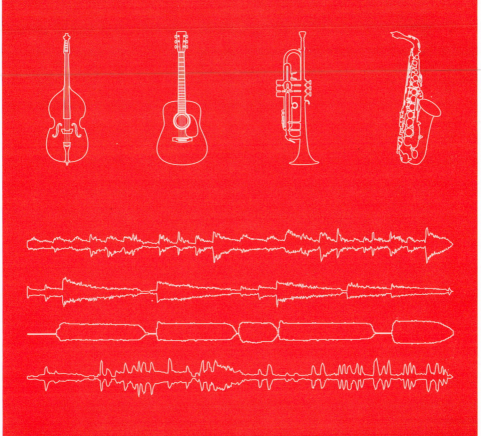

INDEX